Resonant Tunnelling Diode Photonics Devices and Applications (Second Edition)

Online at: https://doi.org/10.1088/978-0-7503-5714-2

Resonant Tunnelling Diode Photonics Devices and Applications (Second Edition)

Charlie Ironside
Department of Physics and Astronomy, Curtin University, Bentley, Australia

Bruno Romeira
International Iberian Nanotechnology Laboratory, Braga, Portugal

José Figueiredo
Faculty of Sciences of the University of Lisbon, Lisbon, Portugal

IOP Publishing, Bristol, UK

ISBN 978-0-7503-5714-2 (ebook)
ISBN 978-0-7503-5712-8 (print)
ISBN 978-0-7503-5715-9 (myPrint)
ISBN 978-0-7503-5713-5 (mobi)

DOI 10.1088/978-0-7503-5714-2

Version: 20231201

IOP ebooks

British Library Cataloguing-in-Publication Data: A catalogue record for this book is available from the British Library.

Published by IOP Publishing, wholly owned by The Institute of Physics, London

IOP Publishing, No.2 The Distillery, Glassfields, Avon Street, Bristol, BS2 0GR, UK

US Office: IOP Publishing, Inc., 190 North Independence Mall West, Suite 601, Philadelphia, PA 19106, USA

This book is dedicated to all our past and existing co-workers who have helped advance resonant tunnelling diode photonics.

Bruno dedicates the book to Raquel, Carolina, Milene, and Adelaide.
José also dedicates the book to Rita, Inês, and Cristina.

Contents

Preface

As semiconductor devices become smaller and smaller, driven by the commercial and technical demands of the semiconductor industry, often summarised as Moore's law that states the number of transistors on a semiconductor chip doubles every 18 months, beyond Moore's law technologies are in urgent need. The semiconductor device dimensions have become small enough that low-energy electrons can tunnel through large energy barriers—an entirely quantum effect. This effect can be utilised to make the highest speed semiconductor devices ever fabricated.

This book is about how a particular version of a quantum device, the resonant tunnelling diode (RTD), is employed in photonic applications. The book generally gives insight into how quantum effects can be used for greater functionality and speed in semiconductor devices. The RTD is a semiconductor device that can act as the highest speed electronic amplifier and oscillator (bandwidth >1 THz). It is made using ultrathin (<10 nm) layers of semiconductor alloy that can be easily integrated with photonic devices such as photodetectors and lasers. Incorporating the ultrahigh-speed RTD amplifier gives the photonic devices new capabilities that are described in the book—as are exciting new applications made possible by integrating the RTD with photonic devices such as neuromorphic (brain-like) photonic computing. Some of these systems provide real-world applications of concepts from nonlinear dynamical theory—concepts including excitability, chaos and synchronisation.

This second edition of the book features a comprehensive introduction to the research literature on resonant tunnelling diode photonics, state-of-the-art introduction into how this technology is deployed to make devices for various systems, and a unique and cogent combination of fundamental physical principles and applications of resonant tunnelling diode photonics. This new edition also features software, Python notebooks, which give the reader not only insight into how existing devices operate, but also tools that can be used to design new devices.

More specifically, this new edition updates the discussion of the devices already presented in the first edition, such as the RTD electroabsorption modulator, the few-photon RTD photodetector, the RTD-driven laser diode, and the RTD optoelectronic oscillator, but also provides new material covering quantum tunnelling, the key physical effect underlying the operation of the RTD, and new devices that can be made with RTDs such as the RTD light-emitting diode and RTD photonic devices for neuromorphic computing. These new additions in the book discuss in more detail how to incorporate these devices into systems that emulate how the brain works, so-called neuromorphic systems, and this requires a detailed understanding of nonlinear dynamics.

In perspective, this second edition of the book uses a combination of quantum theory, semiconductor physics, and nonlinear dynamics to explain how resonant tunnelling diode-based photonic devices can contribute to the development of photonic and neuromorphic systems, extending the book to the area of brain-inspired hardware specifically designed for neural networks which are at the heart of artificial intelligence.

Acknowledgements

We acknowledge the help and support of numerous funders who provided encouragement and resources that helped progress the work reported in this book.

The funders include the Engineering and Physical Science Research Council, (EPSRC), the European Commission research councils, Pond Ventures, and the Fundação para a Ciência e a Tecnologia in Portugal.

We are also pleased to acknowledge our collaborators, including:

Julien Javaloyes, Ignacio Ortega-Piwonka, Edward Wasige, Tony Kelly, Thomas Slight, Robert Hadfield, Weikang Zhang, Scott Watson, Abdullah Al-Khalidi, Bejoys Jacob, Jana Nieder, Qusay Al-Taai, Matej Hejda, Antonio Hurtado, Ekaterina Malysheva, Victor Dolores Calzadilla, Stuart Edwards, J Iwan Davies, Joana Tavares, Gil Cavaco, João Rei, João Lourenço, David Castro Alves, James Foot, Abdullah Al-Khalidi, Jue Wang, Luís Pessoa, Henrique Salgado, Ricardo Avó, Horacio Cantu, Liquan Wang, Ricardo Luz, Joaquim Calado, Isnaldo Coelho, Joaquim Martins Filho.

Author biographies

Charlie Ironside

Professor **Charlie Ironside** is a professor in the Department of Physics and Astronomy at Curtin University (Western Australia). He received his PhD from Heriot-Watt University (Edinburgh) and in 1998 was appointed Professor of Quantum Electronics at the University of Glasgow. He has over 30 years of experience in semiconductor optoelectronics research and in particular, microfabrication of photonic components. He is a Fellow of the IET, a Fellow of the Institute of Physics, and a senior member of IEEE. Along with hundreds of papers, he has co-authored five patents. In 2012 he won the knowledge transfer partnership Scotland award.

Bruno Romeira

Bruno Romeira is a tenured staff researcher at the International Iberian Nanotechnology Laboratory (INL), Portugal. He received a PhD degree (summa cum laude) in physics and the European PhD degree from the University of Algarve, Faro, Portugal, jointly with the University of Glasgow, UK, and the University of Seville, Spain, held a Post-Doctoral Fellowship University of Ottawa, Canada, and then a Marie Skłodowska-Curie Research Fellowship at the Eindhoven University of Technology before joining INL. His research cuts across several disciplines in applied physics and engineering, including semiconductor physics, quantum electronics, low-dimensional nanostructures, nanophotonics, and neuromorphic devices. He leads and participates in several European and national projects in the nanophotonics and neuromorphic fields, including the H2020-FET-OPEN funded project 'ChipAI', the Horizon Europe funded projects 'InsectNeuroNano' and 'CROSSBRAIN', and the FCT-PT funded project 'META-LED'.

Dr Romeira was a recipient of the Young Researchers Incentive Programme Award from the Calouste Gulbenkian Foundation, Portugal, in 2009, the 2011 IEEE Photonics Society Graduate Student Fellowship from the IEEE Photonics Society, USA, and the Best PhD Thesis in optics and photonics in Portugal in 2012 Award from the Portuguese Society of Optics and Photonics. Dr Romeira is a senior member of the Optica Society and of the IEEE Photonics Society.

José Figueiredo

José Figueiredo received a BSc in physics (optics and electronics) and an MSc in optoelectronics and lasers from the Universidade do Porto, Portugal, in 1991 and 1995. He did his PhD work at the University of Glasgow, Scotland, on the optoelectronic properties of resonant tunnelling diodes, enrolled in a joint programme between the Universidade do Porto and the University of Glasgow. Then he moved to the Universidade do Algarve, Portugal. He joined the Physics Department of the Faculty of Sciences of the University of Lisbon in 2017. He is member of the CENTRA research unit (tecnico.ulisboa), Portuguese Society of Optics and Photonics, European Microwave Association, Optical Society of America, and IEEE Photonics Society. In 2001, he co-won the Scottish Proof of Concept Fund for transferring knowledge from university research to industry for commercial exploitation. He co-founded the start-up Essient Photonics, a University of Glasgow spin-out company specialized in optical modulators and photodetectors that take advantage of the properties of resonant tunnelling diodes. He has published over 70 journal and conference papers, and co-authored four patent applications. His research interests include nanoelectronic and optoelectronic devices for terahertz, wireless, radio-over-fibre, neuromorphic, reflectometry, and sensing applications.

IOP Publishing

Resonant Tunnelling Diode Photonics Devices and Applications (Second Edition)

Charlie Ironside, Bruno Romeira and José Figueiredo

Chapter 1

Introduction

This chapter introduces the key concepts governing the design of resonant tunnelling diode (RTD) photonic devices and systems, and puts them in the appropriate context. It covers the electron wavefunction phase-breaking distance and how to determine the mesoscopic physics regime. It also describes quantum tunnelling and nonlinear dynamics, the two main themes of this book. Essentially, RTD photonic devices are an example of optoelectronic integrated circuit (OEIC) technology where photonic and electronic functions are integrated on the same semiconductor chip.

1.1 Introduction

This book combines two broad themes that have generated great interest and excitement in the scientific and technical community over the last 100 years: quantum tunnelling and nonlinear dynamical systems. In particular, it applies them to nanostructured solid-state heterostructures operating at room temperature to gain insight into novel photonic devices, systems, and applications.

During most of the history of microelectronic solid-state devices, charge carriers have been treated as classical particles obeying Newton's laws of classical mechanics—this is the fundamental assumption behind Boltzmann's transport equation and the drift-diffusion model of electronic transport in solid-state devices [1, 2]. How the electron's wave nature interacts with the crystal's period lattice is incorporated into this theory by the effective mass approximation. The electron is treated as a particle that obeys classical mechanics but has a different mass from the free space mass. This semi-classical approximation works because the wave nature of the electron is masked by scattering; that is, the electron collides with entities such as lattice ions, impurity ions and other electrons. These collisions randomly reset the phase of the electron wavefunction. So the electron wavefunction is incoherent on a scale larger than the average distance between collisions, which is called the phase-breaking

doi:10.1088/978-0-7503-5714-2ch1

distance [3]. This phase-breaking distance is larger than the lattice spacing, which is typically around 0.5 nm—which is why the effective mass approximation works. At normal working temperatures, around room temperature, and in typical semiconductors, the phase-breaking distance is less than around 10 nm or so [4]. Another way to express this is that phase breaking is a decoherence effect. If we are to observe, and indeed make use of, quantum mechanical tunnelling effects, then we have to engineer structures that are small enough to preserve the coherence of the electron wavefunction. So, typically, the critical dimensions of quantum tunnelling devices in room temperature semiconductors are approximately 10 nm. This is the scale of the tunnelling devices that we discuss in this book—the so-called mesoscopic scale [4]. To quote Alain Aspect, *'between the scale of a single atom and the macroscopic world, one finds the mesoscopic scale, where it is the object itself, and not only the material of which it is made, that needs to be described by quantum mechanics'* [5].

Using semiconductor growth and nanofabrication tools, we can make mesoscopic devices that utilise the nature of the electron wave to engineer new and exciting charge carrier transport and optical properties, and produce macroscopic, room temperature, and nonlinear current–voltage (I–V) characteristics that exhibit negative differential conductance, i.e., in some part of the I–V curve, as the voltage increases the current decreases. This is achieved by utilising quantum mechanical tunnelling through a resonant state. These are two-terminal devices, and thus are referred to as diodes, so we have an RTD. The RTD [6] is a mesoscopic device, which is deliberately designed to employ the wave nature of the electron to achieve device benefits.

The importance of scale and the transition between the appropriateness of a quantum description of charge carrier transport compared to a classical description is neatly illustrated in [7].

The wave nature of electrons in solids has become increasingly important because the drive to make more compact and faster devices has led to diminishing dimensions of many important features in solid-state electronics. The size of a metal oxide field effect transistor (MOSFET) gate has shrunk to less than 5 nm (at the time of writing in 2022) and it continues to shrink. It can be anticipated that the wave nature of electrons and quantum tunnelling will play an essential part in the MOSFET's operation because the dimensions are approaching the typical phase-breaking distance in room temperature silicon. Tunnelling theory (see, for example, [8]) is already used to describe leakage currents in MOSFET gates that are caused by quantum tunnelling, the so-called direct tunnelling and Fowler–Nordheim tunnelling [9].

The epitaxial growth of semiconductor heterostructures greatly facilitated the feasibility of fabricating mesoscopic devices. Techniques such as molecular beam epitaxy (MBE) and metal organic chemical vapour deposition (MOCVD) can accurately deposit on a substrate crystalline epilayers of semiconductor alloys with thickness as small as one monolayer (~0.5 nm). The GaAs/AlGaAs semiconductor alloy was the first to be investigated and was applied in photonics, leading to the remarkable development of the double heterostructure semiconductor laser [10]. These growth techniques allowed the development of mesoscopic photonic devices that deliberately employed the electron wave nature of the electron to achieve improved functionality, which include the quantum well laser and the quantum cascade laser.

1.2 Quantum tunnelling devices

In terms of semiconductor devices, the so-called Esaki diode [11] was the first device that employed the quantum tunnelling effect to achieve a nonlinear current–voltage curve. The Esaki diode is a *p–n* junction diode with highly doped *p* and *n* regions that give rise to a small depletion width on a scale that allows tunnelling, the depletion width is smaller than the phase-breaking distance. The Esaki diode was the first demonstrated mesoscopic electronic device operating at room temperature, and its discovery resulted in Esaki being awarded the Nobel Prize in 1973. The same year, Josephson was awarded the prize for discovering the superconducting metal–insulator–metal device, the Josephson junction, that relies on quantum tunnelling for its operation [12].

In this book, we are concerned with RTD devices which belong to a class of semiconductor heterostructure devices termed mesoscopic devices. In these devices, the dimensions in the direction of the current flow are smaller than the so-called phase-breaking length [3] of the charge carrier in the device. The phase-breaking length determines the critical dimensions of the RTD device. Phase breaking is caused by the various scattering processes that a charge carrier is subject to in a semiconductor, such as phonon collisions or electron–electron scattering. Phase breaking can be regarded as the distance over which the electron wavefunction remains coherent. Furthermore, if quantum effects are to play an essential role in these devices, then the temperature must be below the Thouless temperature [3, 13], i.e., *the temperature at which thermal averaging introduces an uncertainty of* π *radians in the electron's phase.*

Since we cover electronic mesoscopic structures combined with photonic devices in this book, it is worthwhile to point out some of the striking analogies between optics and quantum devices—wave mechanics is the unifying theme. In optics, the design of thin film devices such as antireflection coatings proceeds via transfer matrix methods [14] and in a similar way these methods can be applied to electronic quantum devices, such as the RTDs [15–18]. In electronic wave mechanics, the electron wavefunction decays as it penetrates through a barrier. The optics analogy is the evanescent wave [19] that is present in the low refractive index part of optical waveguides, and generally in the low refractive index part of optical devices that employ total internal reflection.

1.3 Negative differential conductance

Many physical mechanisms can give rise to negative differential conductance (NDC), and thereby many devices that display NDC characteristics ranging from Gunn diodes [20, 21] to gas discharge tubes [22]. In this book, we are concerned with NDC arising from quantum tunnelling. How quantum tunnelling leads to NDC is discussed in chapter 2. What we mean by NDC is best explained by reference to figure 1.1, which shows a typical nonlinear current–voltage characteristic for an RTD.

When dc is biased to the appropriate voltage, an NDC device exhibits a region of negative conductance (green-shadow region in figure 1.1), i.e., as the voltage is increased, the current decreases. In the NDC region of the I–V curve, the device has electrical gain, i.e., it generates power rather than absorbs power. When biased to the appropriate voltage, in this example around 1.1 V, in figure 1.1, the NDC region has electrical gain. The amount of electrical gain is determined mainly by the ratio of the peak (I_p) to valley currents (I_v).

Note: Since conductance and resistivity have a reciprocal relationship, the resistance of the devices will be negative in the NDC region and often the NDC region is referred to as the negative differential resistance (NDR) region.

The speed at which the device can switch between the peak to valley currents is not apparent from figure 1.1. The RTD can have an enormous gain-bandwidth due to the speed of quantum mechanical tunnelling switching, which is very fast (<1 ps). Electronic oscillators based on RTDs have operated at up to 1.93 THz [24]. Therefore, much of the research interest is centred on RTDs because of their high-speed operation.

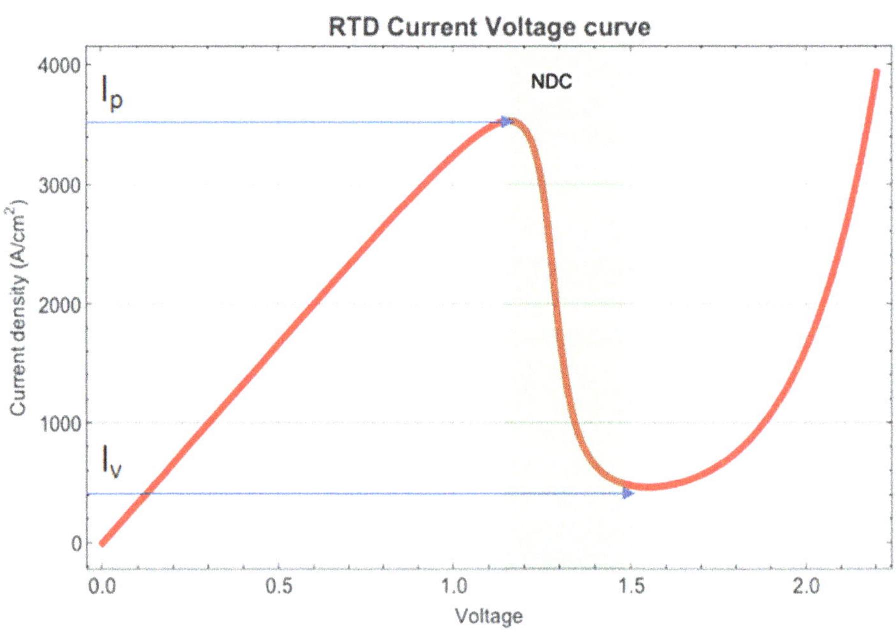

Figure 1.1. The fit to the current–voltage (I–V) curve of a InGaAs/AlAs RTD (taken from [23])—the NDC is exhibited for voltages in the range ~1.1 to 1.5 V—marked on the curve. The peak (I_p) and valley currents (I_v) are shown.

1.4 Nonlinear dynamics

Early in the history of the Esaki tunnel diode, it was realised that the nonlinear *I–V* exhibiting NDC could be used as an analogue of other nonlinear dynamical systems, such as simulating a nerve axon [25], see figure 1.2, which forms part of the FitzHugh–Nagumo model [26] of nerve operation. This model is a simplified version of the well-known Hodgkin–Huxley model describing the behaviour of ion channels that permeate the cell membrane of the squid giant axon. Hodgkin and Huxley were awarded the 1963 Nobel Prize in Physiology or Medicine for this work.

Nagumo *et al* [25], inspired by how the nonlinear operation of nerve cells could be explained by the Bonhoeffer–van der Pol (BVP) model [27], demonstrated how the circuit illustrated in figure 1.2 could model nerve cell activity. Furthermore, they discussed the conditions under which the circuit operated as a Liénard-type oscillator. Much later, the Liénard-type oscillator approach was also applied to the RTD-laser diode (RTD-LD) [28].

To keep in accordance with the Van Der Pol oscillator theory, Nagumo *et al* [25] used the equation of a third-order polynomial equation, $F_N(V)$, to fit the *I–V* characteristics, given by:

$$F_N(V) = i_0 - \frac{1}{\rho}\left((V - V_0) - \frac{(V - V_0)^3}{3K^2}\right) \tag{1.1}$$

where ρ and K are adjustable parameters with the dimensions of resistance and voltage, respectively, and $i_0 = J_0 \times \text{Area}$ and V_0 are shown in figure 1.3.

With this equation (i.e. equation (1.1)), Nagumo *et al* could show the correspondence between the Van der Pol oscillator and the Esaki tunnel diode. Although, as we can see from figure 1.3, the third-order polynomial is not an exact fit for the current–voltage curve that is typically observed with RTDs, it can still be used to describe these oscillators' critical nonlinear dynamic properties (figure 1.4).

Since the original work in the 1960s, an extensive panoply of theoretical tools associated with nonlinear dynamics has been developed—for a general introduction,

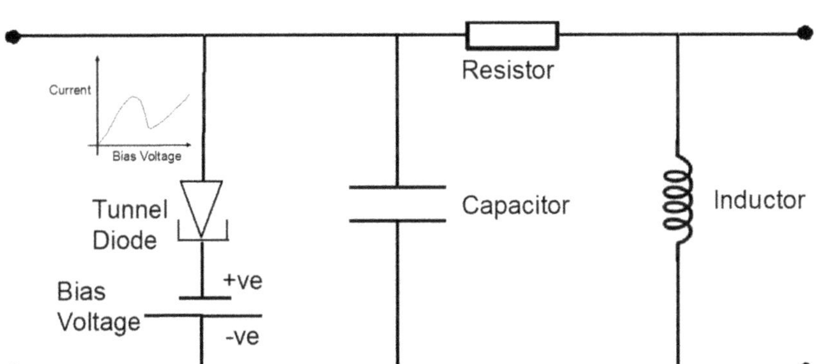

Figure 1.2. An outline circuit diagram of the tunnel-diode nerve model—a tunnel diode with voltage bias, nonlinear *I–V* response and an RCL tank circuit [25].

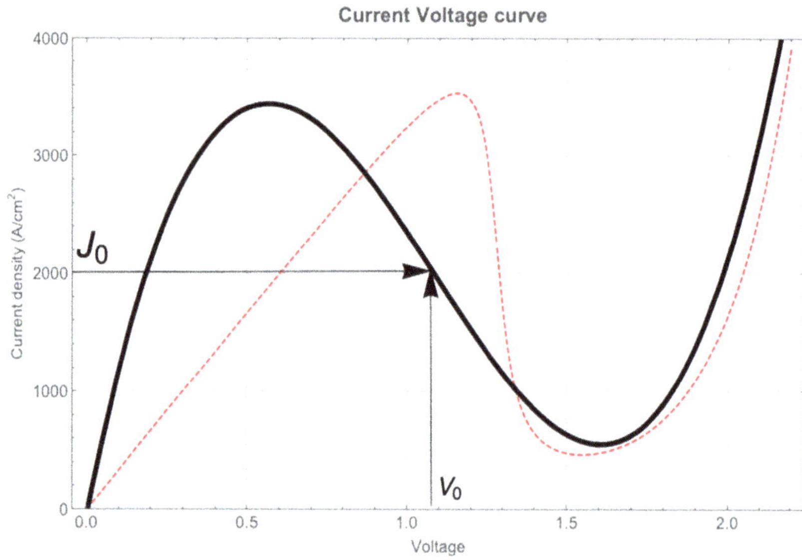

Figure 1.3. **Black line**: The current–voltage curve of a tunnel diode fit with a third-order polynomial equation (1.1); the red dashed line is a comparison with an RTD device (adjustable parameters used to make a comparison $V_0 = 1\text{V}$, $J_0 = 2000\ \text{A cm}^{-2}$, $\rho = 2.4 \times 10^{-4}\ \Omega$ and $K = 0.52$ V).

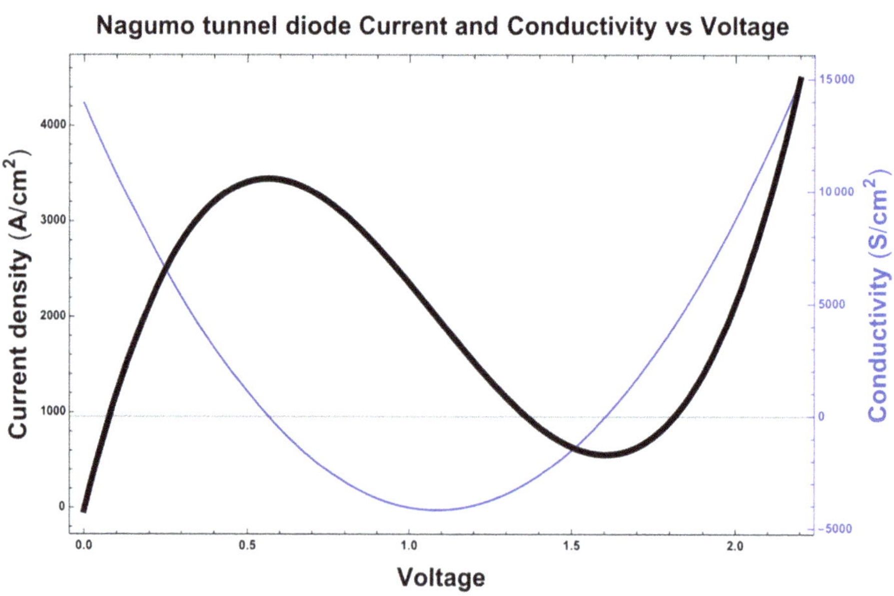

Figure 1.4. Black: The Nagumo (Van Der Pol) model of the tunnel diode current–voltage curve. Blue: The differential w.r.t. voltage of the current–voltage curve to give the conductance, which shows the negative differential conduction (NDC—below zero) region.

see for example [29]. These tools have yielded considerable insight into the operation of both electronic and optoelectronic systems that are dominated by the nonlinear I–V of RTD components, where phenomena such as stochastic resonance [30], chaos, and synchronisation have been observed [31]. Interestingly, the research has now turned full circle and the RTD is being used to build systems that simulate neural computing [32–34].

In this book, we will trace the history of the association between nonlinear dynamics and resonant tunnelling devices, and how it can be used to gain insight into the static and dynamic properties of RTD-based photonic devices and systems.

1.5 Optoelectronic integrated circuits

Optoelectronic technology has been inspired to emulate silicon microfabrication integrated circuits to develop technologies that bring the benefits of micro- and nano-integration to optoelectronic devices and systems. These benefits include high-speed compact devices with greatly enhanced functionality. Two main themes have emerged that are loosely defined as integrated optics, where entirely optical functionality is integrated onto a single chip (e.g. see [35, 36]), and OEICs, where electronic functions such as electronic amplification or logic are combined with optical functions on the same chip (e.g. see [37, 38]).

This book covers how quantum mechanical tunnelling in III–V semiconductors can be employed to achieve OEIC functionality [39] in photonic devices such as modulators (chapter 4), detectors (chapter 5) and lasers and LEDs (chapter 6), and how these new tunnelling OEICs can be combined in innovative systems such as nonlinear dynamic systems (chapter 7) optoelectronic oscillators (chapter 8), and neuromorphic (brain-inspired) photonic computing (chapter 9).

1.6 Outline

The major themes covered in this book and the connections between them are illustrated in figure 1.5. We cover some of the history and background to quantum tunnelling in chapter 2, the electronic and optical properties of RTDs (chapter 3) and their nonlinear dynamic characteristics, emphasising how they are employed to achieve novel RTD photonic devices. This includes resonant tunnelling diode-electroabsorption modulator (RTD-EAM), which is presented in chapter 4; the resonant tunnelling diode-photodetector (RTD-PD), which is presented in chapter 5; and the RTD-LD and resonant tunnelling diode light emitting diode (RTD-LED), which are discussed in chapter 6. The nonlinear dynamic of RTD oscillators is discussed in chapter 7. The resonant tunnelling diode-optoelectronic oscillator (RTD-OEO) is presented in chapter 8. The merging field of RTD photonics for neuromorphic computing is outlined in chapter 9.

Based on the research of many groups worldwide, this book aims to provide a background for the wide range of new opportunities that are emerging for RTD-based photonic devices and systems.

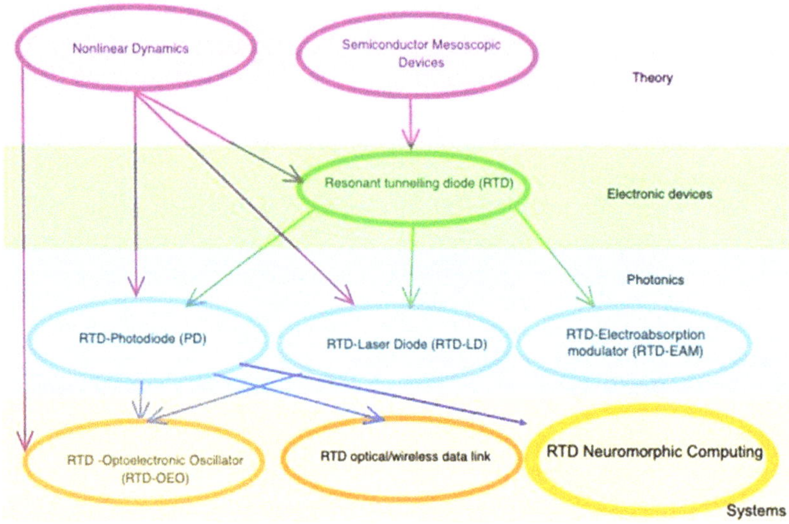

Figure 1.5. The RTD photonics map showing the relationships between theory, electron devices, photonics and RTD photonics-based systems.

References

[1] Sverdlov V, Ungersboeck E, Kosina H and Selberherr S 2008 Current transport models for nanoscale semiconductor devices *Mater. Sci. Eng. R-Rep.* Review **58** 228–70
[2] Nanohub *NEMO—1D* (https://nanohub.org/groups/inemo/software-projects/nemo1d/)
[3] Bandyopadhyay S 2012 *Physics of Nanostructured Solid State Devices* (Berlin: Springer)
[4] Ferry D K 2015 *Transport in Semiconductor Mesoscopic Devices* (Bristol: IOP Publishing)
[5] Bell J S 2004 *Speakable and Unspeakable in Quantum Mechanics: Collected Papers on Quantum Philosophy* 2nd edn (Cambridge: Cambridge University Press)
[6] Mizuta H and Tanoue T 1995 *The Physics and Applications of Resonant Tunnelling Diodes (Cambridge Studies in Semiconductor Physics and Microelectronic Engineering)* (Cambridge: Cambridge University Press)
[7] Granot E 2006 Resonant tunnelling and the transition between quantum and classical domains *Eur. J. Phys.* **27** 985–93
[8] Ranuarez J C, Deen M J and Chen C H 2006 A review of gate tunneling current in MOS devices *Microelectron. Reliab.* **46** 1939–56
[9] Fowler R H and Nordheim L W 1928 Electron emission in intense electric fields *Proc. R. Soc. London, Ser.* A **119** 173
[10] Orton J 2004 *The Story of Semiconductors* (Oxford: Oxford University Press)
[11] Esaki L 2015 What did I explore in half a century of research? What discovery, what invention, where, when? *Jpn. J. Appl. Phys. Rev.* **54** 040101
[12] Luryi S 1985 Frequency limits of double barrier resonant tunneling oscillators *Appl. Phys. Lett.* **47** 490–2
[13] Edwards J T and Thouless D J 1972 Numerical studies of localization in disordered systems *J. Phys. Part C Solid State Phys.* **5** 807
[14] Gerrad A and Burch J M 1975 *Introduction to Matrix Methods in Optics* (New York: Dover)

[15] Price P J 1989 Simple theory of double-barrier tunneling *IEEE Trans. Electron Devices* **36** 2340–3

[16] Hutchings D C 1989 Transfer matrix approach to the analysis of an arbitrary quantum well structure in an electric field *Appl. Phys. Lett.* **55** 1082–4

[17] Walker J S and Gathright J 1992 A transfer-matrix approach to one-dimensional quantum mechanics using Mathematica *Comput. Phys.* **6** 393

[18] Walker J S and Gathright J 1994 Exploring one-dimensional quantum-mechanics with transfer-matrices *Am. J. Phys.* **62** 408–22

[19] Enders A and Nimtz G 1993 Evanescent mode propagation and quantum tunneling *Phys. Rev.* E **48** 632–4

[20] Kroemer H 1968 Negative conductance in semiconductors *IEEE Spectr.* **5** 47–56

[21] Khalid A *et al* 2014 Terahertz oscillations in an In0.53Ga0.47As submicron planar Gunn diode *J. Appl. Phys.* **115** 114502

[22] Mahamud R and Farouk T I 2016 Ion kinetics and self pulsing in DC microplasma discharges at atmospheric and higher pressure *J. Phys. D: Appl. Phys.* **49** 145202

[23] Schulman J N, Santos H J D and Chow D H 1996 Physics-based RTD current–voltage equation *IEEE Electron Device Lett.* **17** 220–2

[24] Maekawa T, Kanaya H, Suzuki S and Asada M 2016 Oscillation up to 1.92 THz in resonant tunneling diode by reduced conduction loss *Appl. Phys. Express* **9** 024101

[25] Nagumo J, Arimoto S and Yoshizawa S 1962 An active pulse transmission line simulating nerve axon *Proc. IRE* **50** 2061–70

[26] FitzHugh R 1960 Thresholds and plateaus in the Hodgkin-Huxley nerve equations *J. Gen. Physiol.* **43** 867–96

[27] Bonhoeffer K F 1948 Activation of passive iron as a model for the excitation of nerve *J. Gen. Physiol.* **32** 69–91

[28] Slight T J, Romeira B, Wang L Q, Figueiredo J M L, Wasige E and Ironside C N 2008 A lienard oscillator resonant tunnelling diode-laser diode hybrid integrated circuit: model and experiment *IEEE J. Quantum Electron.* **44** 1158–63

[29] Lynch S 2017 *Dynamical Systems with Applications Using Mathematica* (Birkhäuser: Springer)

[30] Hartmann F, Gammaitoni L, Hofling S, Forchel A and Worschech L 2011 Light-induced stochastic resonance in a nanoscale resonant-tunneling diode *Appl. Phys. Lett.* **98** 242109

[31] Romeira B *et al* 2008 Synchronisation and chaos in a laser diode driven by a resonant tunnelling diode *IET Optelectron.* **2** 211–5

[32] Romeira B, Figueiredo J M L and Javaloyes J 2017 Delay dynamics of neuromorphic optoelectronic nanoscale resonators: perspectives and applications *Chaos* **27** 114323

[33] Romeira B, Avo R, Javaloyes J, Balle S, Ironside C N and Figueiredo J M L 2014 Stochastic induced dynamics in neuromorphic optoelectronic oscillators *Opt. Quantum Electron.* **46** 1391–6

[34] Klofai Y, Essimbi B Z and Jager D 2015 An MMIC implementation of FitzHugh–Nagumo neurons using a resonant tunneling diode nonlinear transmission line *Phys. Scr.* **90** 025002

[35] Tamir T, Giora G and Bertoni H L 1994 Guided-Wave Optoelectronics: Device Characterization, Analysis, and Design *Int. Symp. Proc. Held in Brooklyn, NY on October 26–28, 1994*

[36] Tamir T (ed) 1975 *Integrated Optics* (Series Topics in Applied Physics) (Berlin: Springer)

[37] Atef M and Zimmermann H 2016 *Optoelectronic Circuits in Nanometer CMOS Technology* (Springer Series in Advanced Microelectronics) (Berlin: Springer)
[38] Gao J 2011 *Optoelectronic Integrated Circuit Design and Device Modeling* (Wiley)
[39] Ironside C N, Figueiredo J M L, Romeira B, Slight T J, Wang L and Wasige E 2010 Resonant tunneling diode optoelectronic integrated circuits *Quantum Sens. Nanophoton. Dev.* **Vii** 76080i

IOP Publishing

Resonant Tunnelling Diode Photonics Devices and Applications (Second Edition)

Charlie Ironside, Bruno Romeira and José Figueiredo

Chapter 2

Quantum tunnelling

This chapter discusses quantum tunnelling, which is the key physical effect underlying the operation of the resonant tunnelling diode (RTD). Quantum tunnelling is essential in many areas of physics and engineering. In addition, in broad outline, this chapter covers how quantum tunnelling explains RTD negative differential conductance (NDC). The objectives are to explain how the NDC of the RTD arises and to give insight into why the RTD has an extremely fast response time.

2.1 Introduction

One of the early successes of quantum theory was the use of quantum tunnelling to give insights into alpha particle radioactive decay [1–3]. The mystery of alpha particle decay, for which classical physics had no explanation, was that the alpha particles are emitted from the nucleus with less energy than the height of the energy barriers that confine the particles to the nucleus. Subsequently, insights provided by quantum tunnelling theory have been applied in many areas of physics, engineering, chemistry [4], astronomy [5] and biology [6].

A low-energy particle surmounts relatively high energy barriers due to the dual wave-particle nature of matter. This concept was first introduced by DeBroglie [7] to understand the resonant or bound atomic states that had been inferred from atomic spectroscopy. Schrödinger developed a wave equation for matter waves. The dual particle-wave nature of matter means that the square of its wavefunction gives the position of an alpha particle (or electron), and the wavefunction is a solution to Schrödinger's equation. Energy barriers set up the boundary conditions for solving Schrödinger's equation. For finite energy barriers, there is always a finite amount of the wavefunction outside the energy barrier. Therefore, there is always a finite probability of finding the particle outside of the energy barrier, and that probability diminishes as the barrier height increases. Thus, the particle can come through (tunnel through) the barrier rather than over the barrier. Furthermore, as we shall

doi:10.1088/978-0-7503-5714-2ch2

see with RTDs, introducing resonance states can enhance the probability of tunnelling through the barrier.

2.2 Negative differential resistance

Here we give a broad outline of a 'first order' theory to account for how quantum tunnelling leads to negative differential conductance (resistance). To keep it simple, this account leaves out many processes that complicate electron wavefunction propagation in semiconductors. For example, it neglects thermal and impurity scattering processes that lead to the decoherence of the electron wavefunction. For a more realistic and complete account, the reader is referred to more sophisticated models on the Nanohub website [8–10].

The resonance state of the RTD gives the wavefunction of the incoming electron as an additional quantum pathway or channel through the RTD. How this leads to the RTD exhibiting negative differential resistance (NDR) is described here. The resonance state provides a transmission channel because of the overlap between the wavefunction of the incoming electron and the resonance state wavefunction. The degree of overlap determines the coupling between the initial and final states, as described by Fermi's golden rule [11]. The resonance state channel is only available to incoming electrons with a wavefunction that 'fits' the resonance state wavefunction, i.e., the wavevector, k, must produce a wave that has an integer number of half wavelengths, $\lambda = \frac{2\pi}{k}$, that is equal to the distance between the barriers of the RTD. In the emitter region of the RTD, the electrons will have a spectrum of wavevectors, and the resonance channel will select a small range of this spectrum. The wavevector is related to the energy of the electron, $E = \frac{\hbar^2 k^2}{2m}$. As the voltage, V, applied to the RTD changes, the energy of incoming electrons and thus the wavevector of the incoming electrons can be tuned. The maximum current through the RTD is obtained when the peak of the electron wavevector spectrum fits the resonance state channel. If the voltage is increased beyond this point, then there is a dramatic drop in the current. This drop results from the nature of the distribution curve of the electron spectrum, which is a Fermi distribution that drops off rapidly to zero for low energies. The current drops as the voltage is increased beyond maximum current, and thus NDR.

Much of the above paragraph is succinctly expressed in the following equation from Esaki and Tsu [12, 13]

$$J(V) = \frac{em \times kT}{\pi^2 \hbar^3} \int_0^\infty T(E, V) \ln\left[\frac{1 + \exp(E_F - E)/kT}{1 + \exp(E_F - E - eV)/kT}\right] dE \tag{2.1}$$

where $J(V)$ is the current density at voltage, V; e is the charge on the electron; k is Boltzman's constant; T is temperature; $\hbar$ is the reduced Planck constant; $T(E, V)$ is the transmission probability; and E_F is the Fermi energy of the electron energy distribution in the emitter.

By solving Schrödinger's equation with the appropriate energy barrier configuration, it is possible to accurately estimate the probability of tunnelling through energy barriers, $T(E, V)$. In the next section, we show how the transfer matrix method can calculate the tunnelling probability versus electron energy and how it can be applied to an AlGaAs/GaAs/AlGaAs RTD.

2.3 Calculating the transmission probability, $T(E)$

The transfer matrix method is used to deal with waves interacting with various structures in one dimension, which has been extensively employed because it relatively quickly yields numerical solutions and physical insight [14, 15]. It has been used to model the closely related metal insulator metal (MIM) diodes [16] and RTDs. Here we will use it to calculate the transmission probability through an RTD as a function of electron energy [17], $T(E)$.

2.3.1 $T(E)$, via the transfer matrix method

We start with the one-dimensional and time-independent version of Schrödinger's equation:

$$-\frac{\hbar^2}{2m}\frac{d^2\Psi(x)}{dx^2} = [E - V(x)]\Psi(x) \tag{2.2}$$

where $\hbar$ is the reduced Planck's constant, m is the mass of the electron, x is the spatial dimension perpendicular to the plane of the epitaxial layers, E is the kinetic energy of the electron and $V(x)$ is the spatial varying potential shown in figure 2.1(a). This spatially varying potential is achieved using semiconductor alloy epilayers with varying energy bandgap.

Solutions for the electron wavefunction, $\Psi(x)$, are in the form of plane waves moving to the right Ae^{ikx} (on figure 2.1(a)) and to the left Be^{-ikx}. So, for example, at the first barrier, A is the fraction of wavefunction reflected from barriers and B is the fraction of the wavefunction reflected, and due to the conservation of charge $A + B = 1$.

$$\Psi(x) = Ae^{ikx} + Be^{-ikx} \tag{2.3}$$

where the electron wavevector, $k = \frac{\sqrt{2mV(x)}}{\hbar}\sqrt{\frac{E}{V(x)} - 1}$.

In matrix form:

$$\Psi(x) = (e^{ikx}e^{-ikx})\begin{pmatrix} A \\ B \end{pmatrix} \tag{2.4}$$

As illustrated in figure 2.1(b), each part of the RTD is represented by a matrix that describes how the part changes the electron wavefunction.

So, for example, the emitter region, of length, L_0, is described by a propagation matrix:

$$\boldsymbol{P}_0 = \begin{bmatrix} e^{ik_0L_0} & 1 \\ 0 & e^{-ik_0L_0} \end{bmatrix} \tag{2.5}$$

where $k_0 = \frac{\sqrt{2m}}{\hbar}\sqrt{E - V_0}$.

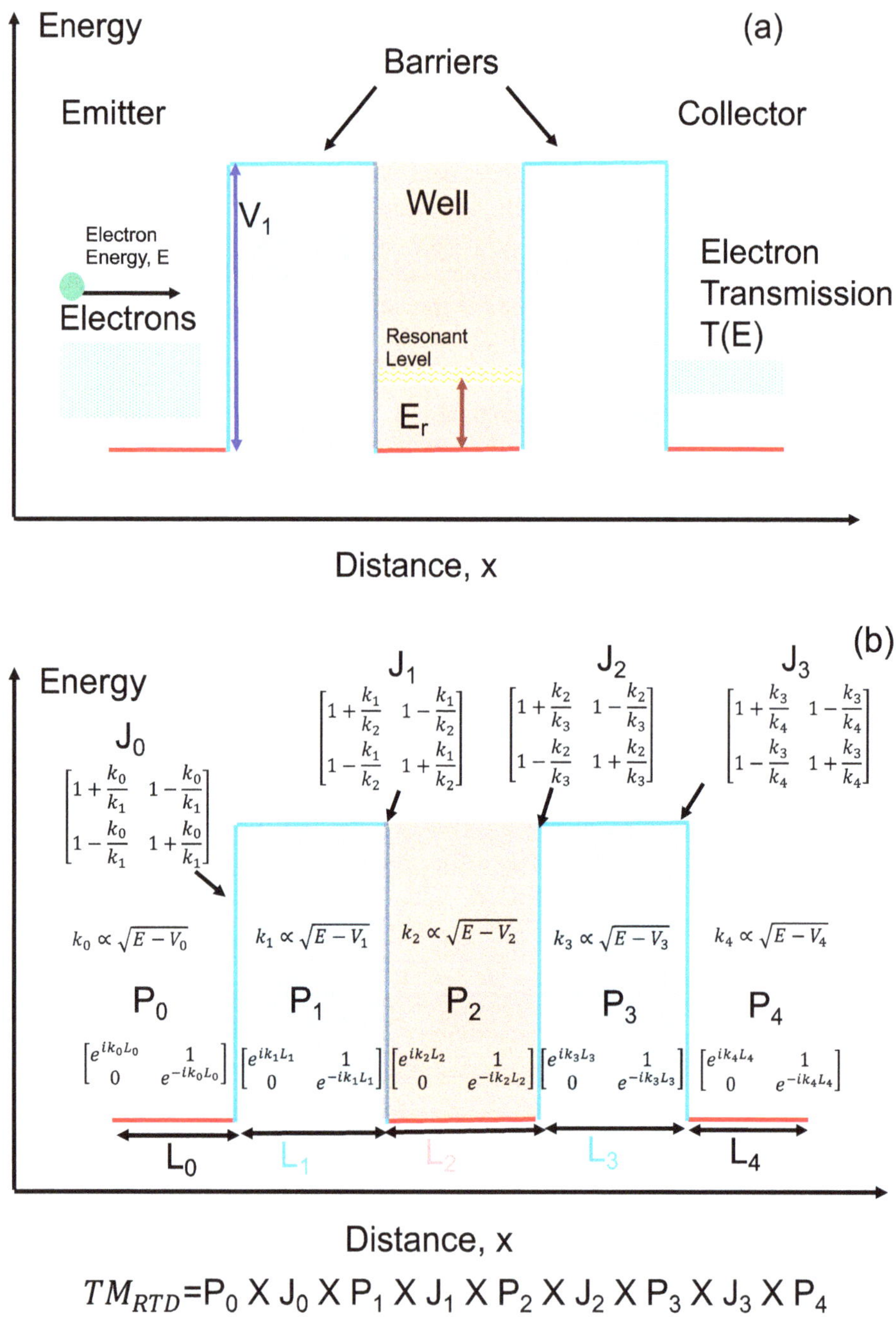

Figure 2.1. (a) The generic semiconductor RTD energy band diagram showing the bottom of the conduction band energy versus distance; the barriers are formed by a relatively larger bandgap semiconductor alloy. Panel (b) shows how the RTD transfer matrix (TM_{RTD}) is calculated.

The first energy step associated with the first barrier is described by an energy jump matrix:

$$J_0 = \begin{bmatrix} 1 + \dfrac{k_0}{k_1} & 1 - \dfrac{k_0}{k_1} \\ 1 - \dfrac{k_0}{k_1} & 1 + \dfrac{k_0}{k_1} \end{bmatrix} \tag{2.6}$$

where $k_1 = \frac{\sqrt{2m}}{\hbar}\sqrt{E - V_1}$.

The overall transmission matrix representing the RTD, TM_{RTD}, is found by multiplying the matrix for each element of the RTD, as follows:

$$TM_{\mathrm{RTD}} = P_0 \times J_0 \times P_1 \times J_1 \times P_2 \times J_2 \times P_3 \times J_3 \times P_4 \tag{2.7}$$

If we write the matrix elements of the TM_{RTD} matrix, as follows:

$$TM_{\mathrm{RTD}} = \begin{bmatrix} M_{11} & M_{12} \\ M_{21} & M_{22} \end{bmatrix} \tag{2.8}$$

The transmission, $T(E)$, is then given by [18]:

$$T(E) = 1 - \frac{|M_{21}|^2}{|M_{11}|^2} \tag{2.9}$$

Thus, the energy dependent transmission coefficient $T(E)$ of the electron wavefunction can be determined from elements in the matrix that represents the RTD.

To determine the transmission of the RTD structure for a range of electron energies, $T(E)$, the transfer matrix representing the RTD is determined for each energy in an array spanning the energy range of interest.

2.4 Applying the transfer matrix method to an AlGaAs/GaAs/AlGaAs RTD

The method described in section 2.4.1 can easily be converted into an algorithm implemented in software. A Python notebook has been produced to illustrate this method in action (see chapter 11). This method has been applied to an AlGaAs/GaAs/AlGaAs RTD and the results are presented in this section.

2.4.1 The $Al_xGa_{1-x}As$ layers

Chapter 2 of [19] gives the essential materials parameters for $Al_xGa_{1-x}As$ (see also [20]). The device is illustrated in figure 2.2.

To calculate the transfer matrix associated with a given epilayer, both the energy of the conduction band discontinuity and the effective mass are required to calculate the electron wavevector k in the epilayer.

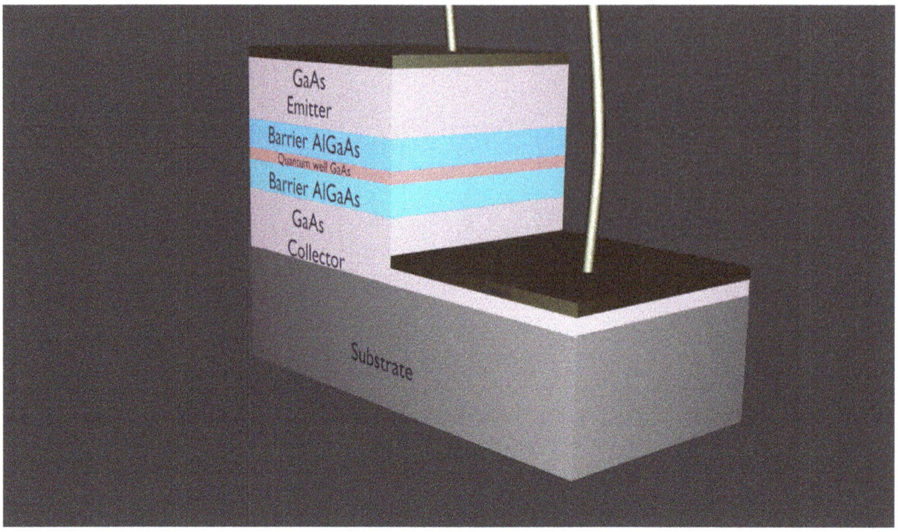

Figure 2.2. Illustration of a generic $Al_xGa_{1-x}As/GaAs/Al_xGa_{1-x}As$ RTD device. Showing the epitaxial layers and electrical contacts to the device.

The conduction band discontinuity as a function of the Al fraction, x, (unfortunately, it is the usual convention to use the symbol 'x' to represent the spatial dimension and the Al fraction—here we use x in italics to refer to the Al fraction) is given by [19]:

$$\begin{aligned} &V = \beta \times 1.247\,x \text{ for } (0 < x < 0.45) \\ &\text{otherwise } \Delta E = \beta \times (1.247\,x + 1.147(x - 0.45)^2) \end{aligned} \tag{2.10}$$

where $\beta = 0.6$ and the V is given in eV.

And the effective mass, m^*, as a function of Al fraction, x:

$$m^* = (0.067 + 0.081x)m_o \tag{2.11}$$

where m_o is the mass of the free electron mass.

Figure 2.3(a) shows the energy band structure of an $Al_xGa_{1-x}As/GaAs/Al_xGa_{1-x}As$ RTD and the transmission probability, $T(E)$, calculated using the transfer matrix method as implemented using the Python notebook (section 11.1.2) run on Google's Colab platform [21]. Figure 2.3(b) clearly shows a resonance peak at energies below the height of the barriers.

If we compare figure 2.3(b) and figure 2.4(b), then the role of the two barriers becomes apparent. In figure 2.4, the second barrier is removed, and therefore there are no resonance peaks for energies below the barrier height. There is no quantum pathway, no resonance channel, available for the electrons in the single barrier structure. This leads to the somewhat counter-intuitive conclusion that adding an energy barrier increases the probability of transmission through the structure at energies below the height of the barrier.

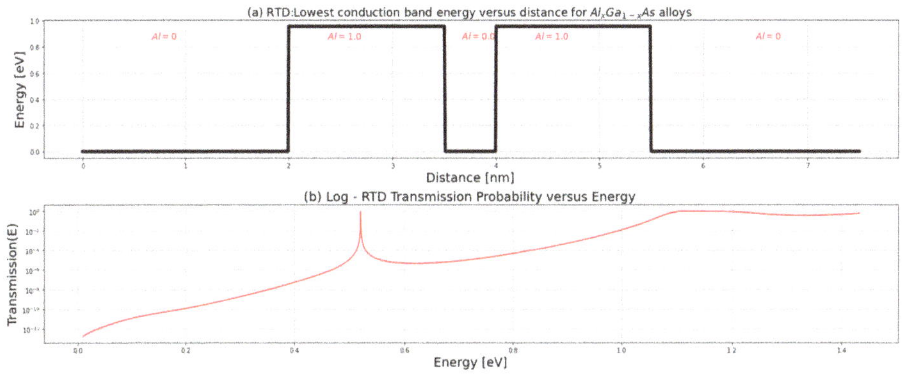

Figure 2.3. (a) The energy band structure of an RTD—the bottom of the conduction band versus distance—with 1st Barrier width, B1 =1.50 nm 1st Barrier Al fraction, $x = 1.0$, 1st Quantum Well width, QW = 0.5 nm, QW Al fraction, $x = 1.0$, 2nd Barrier width = 1.5 nm width, B1 = 1.50 nm 2nd Barrier Al = 0.4; (b) is a semilog plot of the electron transmission probability, $T(E)$ versus energy calculated using the transfer matrix method.

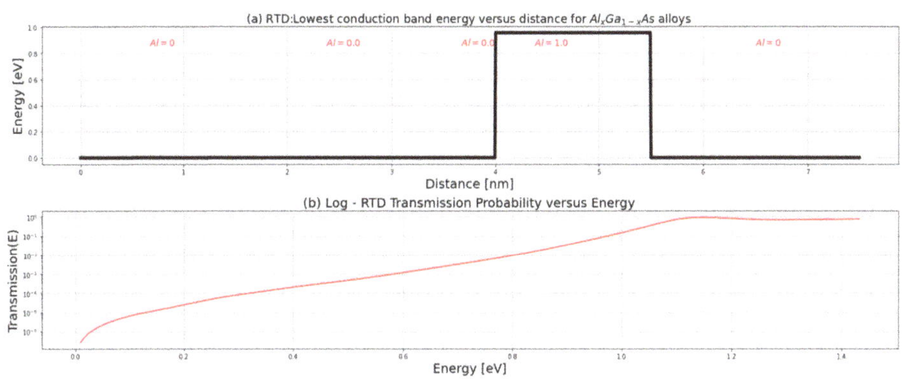

Figure 2.4. (a) The energy band structure of a barrier with 2nd Barrier width, B2 = 1.50 nm Barrier Al fraction, $x = 1.0$, there is no quantum well or first barrier; (b) is a semilog plot of the electron transmission probability, $T(E)$, versus energy.

2.5 Resonance tunnelling diode response time

The response time of the RTD is a crucial parameter. The inverse of the response time, the enormous bandwidth of RTD-based devices and systems, has been and continues to be an essential driver of their development. Here we build on the transfer matrix method and give a brief straightforward introduction to the topic that emphasises the quasi-bound state lifetime.

The RTD acts like a resonator for the electron wavefunction that produces a quasi-bound state. As with all resonators, there is a storage time or effective transit time associated with the resonance (also called the quasi-bound-state lifetime [22]), which limits the frequency response of the RTD. The longer the storage time, the higher the Q (or Quality of the resonator) and the smaller the response bandwidth. An upper limit to the response time can be estimated from the first (lowest energy) resonance peak in the $T(E)$ versus energy, shown in figure 2.3. If the peak is examined in more detail, then the width of the peak, w, can be measured. The quasi-bound-state lifetime of the RTD is then approximately [23, 24], $\tau \approx h/w$ and the maximum bandwidth [25] $f_m \approx 1/(2\pi\tau)$.

It must be noted that the response time, τ, deduced from the transfer matrix method is an upper limit—the transfer matrix method takes no account of scattering effects that will spoil the Q of the RTD resonance, and it takes no account of the applied electric field, external circuit or any nonlinear response of the RTD [26]. With these caveats in mind, we can obtain an inherent upper limit to the response time, and thus a lower limit to the bandwidth of an RTD. For a much more sophisticated model of the transient response time of an $Al_xGa_{1-x}As/GaAs/Al_xGa_{1-x}As$ RTD that includes scattering induced incoherence effects on the electron wavefunction, see [27].

Figure 2.5 shows a linear plot of the first (lowest energy) peak from figure 2.3(b) with an expanded energy scale fitted to a Lorentzian curve [28] with a fullwidth at half-height, $w = 8.93 \times 10^{-5}$ eV—so from $\tau \approx h/w$ and Planck's constant $h = 6.58 \times 10^{-16}$ eV s, $\tau \approx 7$ ps. Remember that this is an upper limit. The analysis of [27] that includes scattering reduces the value of $\tau \approx 0.1$ ps.

Although the transfer matrix method only gives an upper limit to the response time of the RTD, it does provide some insight into the trade-offs involved in optimising the RTD for a given application. For example, if the barrier heights are lowered, then the response time will be reduced, thereby increasing the RTD bandwidth, but other channels through the RTD become enhanced. Lowering barriers results in the incoming electrons having a higher probability of directly surmounting the barriers rather than using the resonance state channel. The current–voltage curve will show a reduced peak-to-valley current, which is a classic gain-bandwidth trade-off. Thus, all else held constant, we can expect that as the RTD oscillators increase in operating frequency, the output power will decrease.

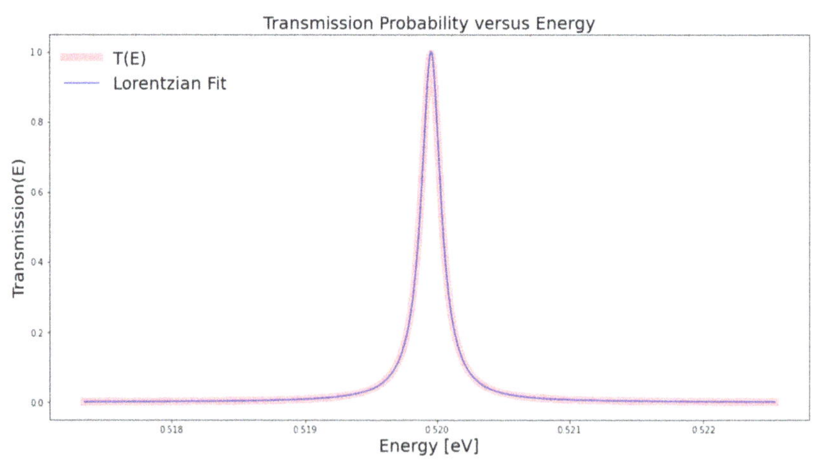

Figure 2.5. A linear plot of the first (lowest energy) peak from figure 2.3(b) with an expanded energy scale fitted to a Lorentzian curve with a fullwidth at half-height, $w = 8.9 \times 10^{-5}$ eV.

However, if we compare the transfer matrix method with experimental measurements of the RTD transfer times, the incompleteness of the theory becomes apparent. Measurements of transfer times through quantum wells [29, 30] indicate that tunnelling times decrease as the barrier thickness increases, and the transfer matrix method predicts that the tunnelling times increase as the barrier thickness increases. This discrepancy happens because the transfer matrix method, among other things, ignores scattering processes in the barriers. The transfer matrix method predicts that with increasing barrier width, the barrier reflects more of the electron wavefunction, thus increasing the Q of the RTD, and thereby increasing the quasi-bound lifetime. However, as observed in [27], increased scattering in the thicker barrier reduces the reflectivity and the quasi-bound lifetime. More complete modelling [8, 31] includes scattering.

For a more detailed analysis of how the quasi-bound-state lifetime, τ, is taken into account in circuit design and analysis as a quantum well inductance and employed to design a $Al_xGa_{1-x}As/GaAs/Al_xGa_{1-x}As$ RTD, 420 GHz, oscillator, see [22]. The relationship between the quasi-bound-state lifetime, τ, and high-frequency performance of an RTD-based device is also discussed in [32], and see [33] for a recent update on large signal equivalent circuits for RTD-based systems that takes the nonlinear effects inherent to RTDs into account.

In some tunnelling devices, the response can be so fast that they can be employed as a rectifier for light, i.e., even at optical frequencies, tunnelling devices can convert AC from light into DC that can be used to drive external circuits. MIM diodes are closed related to RTD and can be analysed using the transfer matrix method [34]. MIMs are lower inductance and resistance than RTDs, and respond more quickly. There are several proposals [35–37] to use MIMs in combination with nanoscale antennae to harvest solar energy as alternatives to photovoltaics. In principle, this approach is not constrained by the Shockley–Queisser [38] limit on conversation efficiency, which is around 25% for single junction silicon-based photovoltaics.

2.6 Conclusions

In this chapter, we have briefly discussed the inception and history of quantum mechanical tunnelling theory, and how it has been employed to give insight into many physical phenomena. We showed how a particular version of the tunnelling theory, the transfer matrix method, can be applied to an AlGaAs/GaAs/AlGaAs RTD to determine the transmission coefficient versus energy. The theory is relatively simple to use and code. It also gives some physical insight into the operation of the RTD, including providing an upper bound to the response time of the RTD. However, it is a relatively incomplete model and more sophisticated models are required to accurately predict the operation of RTDs.

References

[1] Merzbacher E 2002 The early history of quantum tunnelling *Phys. Today* **55** 44–9

[2] Gamow G 1928 The quantum theory of nuclear disintegration *Nature* **122** 805–6

[3] Le Deunff J, Brodier O and Mouchet A 2012 A primer for resonant tunnelling *Eur. J. Phys.* **33** 1771–87
[4] McMahon R J 2003 Chemical reactions involving quantum tunneling *Sci. Editor. Mater.* **299** 833–4
[5] Hiraoka K, Sato T and Takayama T 2001 Interstellar chemistry—tunneling reactions in interstellar ices *Sci., Editor. Mater.* **292** 869–70
[6] Sutcliffe M J and Scrutton N S 2000 Enzyme catalysis: over-the-barrier or through-the-barrier? *Trends Biochem. Sci.* **25** 405–8
[7] de Broglie L 2021 *Research on the Theory of Quanta* (Minkowski Institute Press)
[8] Park H-H *et al* 2022 Resonant Tunneling Diode Simulation with NEGF. (https://nanohub.org/resources/rtdnegf) (accessed August 2022)
[9] Klimeck G 2022 *RTD with NEGF Demonstration: Basic RTD Asymmetric* (https://nanohub.org/resources/6812) (accessed August, 2022)
[10] Klimeck G, Lake R, Bowen R C, Frensley W R and Moise T S 1995 Quantum device simulation with a generalized tunneling formula *Appl. Phys. Lett.* **67** 2539–41
[11] Reittu H J 1995 Fermi's golden rule and Bardeen's tunneling theory *Am. J. Phys.* **63** 940–4
[12] Tsu R and Esaki L 1973 Tunneling in a finite superlattice *Appl. Phys. Lett.* **22** 562–4
[13] Schulman J N, Santos H J D and Chow D H 1996 Physics-based RTD current–voltage equation *IEEE Electron Device Lett.* **17** 220–2
[14] Sanchez-Soto L L, Monzon J J, Barriuso A G and Carinena J F 2012 The transfer matrix: a geometrical perspective *Phys. Rep.-Rev. Sect. Phys. Lett.* **513** 191–227
[15] Pereyra P 2022 The transfer matrix method and the theory of finite periodic systems. from heterostructures to superlattices *Phys. Status Solidi B-Basic Solid State Phys.* **259** 2100405
[16] Abdolkader T M, Shaker A and Alahmadi A N M 2018 Numerical simulation of tunneling through arbitrary potential barriers applied on MIM and MIIM rectenna diodes *Eur. J. Phys.* **39** 045402
[17] Walker J S and Gathright J 1994 Exploring one-dimensional quantum-mechanics with transfer-matrices *Am. J. Phys.* **62** 408–22
[18] Miller D A B 2008 *Quantum Mechanics for Scientists and Engineers* (Cambridge: Cambridge University Press)
[19] Mizuta H and Tanoue T 1995 *The Physics and Applications of Resonant Tunnelling Diodes (Cambridge Studies in Semiconductor Physics and Microelectronic Engineering)* (Cambridge: Cambridge University Press)
[20] AlGaAs Alloy (http://www.ioffe.ru/SVA/NSM/Semicond/AlGaAs/index.html) (accessed 2022)
[21] Colab Google (https://colab.research.google.com) (accessed 2022)
[22] Brown E R, Sollner T, Parker C D, Goodhue W D and Chen C L 1989 Oscillations up to 420 GHz in GaAs/AlAs resonant tunneling diodes *Appl. Phys. Lett.* **55** 1777–9
[23] Ricco B and Azbel M Y 1984 Physics of resonant tunneling—the one dimensional double barrier case *Phys. Rev.* B **29** 1970–81
[24] Anemogiannis E, Glytsis E N and Gaylord T K 1993 Bound and quasi-bound state calculations for biased/unbiased semiconductor quantum heterostructures *IEEE J. Quantum Electron.* **29** 2731–40
[25] Luryi S 1985 Frequency limits of double barrier resonant tunneling oscillators *Appl. Phys. Lett.* **47** 490–2

[26] Villani M, Clochiatti S, Prost W, Weimann N and Oriols X 2021 There is plenty of room for THz tunneling electron devices beyond the transit time limit *IEEE Electron Device Lett.* **42** 224–7

[27] Frensley W R 1987 Wigner-function model of a resonant tunneling semiconductor device *Phys. Rev.B* **36** 1570–80

[28] Price P J 1986 Resonant tunneling properties of heterostructures *Superlattices Microstruct.* **2** 593–6

[29] Alexander M G W, Nido M, Ruhle W W and Kohler K 1990 Resonant-tunneling transfer times between asymmetric GaAs Al0.35Ga0.65 double quantum wells *Phys. Rev.* B **41** 12295–8

[30] Oberli D Y *et al* 1989 Direct measurement of resonant and nonresonant tunneling times in asymmetric coupled quantum wells *Phys. Rev.* B **40** 3028–31

[31] Cimbri D *et al* 2022 Accurate quantum transport modeling of high-speed $In_{0.53}Ga_{0.47}As$/AlAs double-barrier resonant tunneling diodes *IEEE Trans. Electron Devices* **69** 4638–45

[32] Feiginov M N 2000 Does the quasibound-state lifetime restrict the high-frequency operation of resonant-tunnelling diodes? *Nanotechnology* **11** 359–64

[33] Ourednik P, Picco G, Nguyen D T, Spudat C and Feiginov M 2023 Large-signal dynamics of resonant-tunneling diodes *J. Appl. Phys.* **133** 014501

[34] Grover S and Moddel G 2012 Engineering the current–voltage characteristics of metal–insulator–metal diodes using double-insulator tunnel barriers *Solid-State Electron.* **67** 94–9

[35] Hamied F M A, Mahmoud K R, Hussein M and Obayya S S A 2022 Design and analysis of a nano-rectenna based on multi-insulator tunnel barrier for solar energy harvesting *Opt. Quantum Electron.* **54** 144

[36] Jayaswal G, Belkadi A, Meredov A, Pelz B, Moddel G and Shamim A 2018 Optical rectification through an Al2O3 based MIM passive rectenna at 28.3 THz *Mater. Today Energy* **7** 1–9

[37] Kale B M 1985 Electron-tunneling devices in optics *Opt. Eng.* **24** 267–74

[38] Shockley W and Queisser H J 1961 Detailed balance limit of efficiency of p–n junction solar cells *J. Appl. Phys.* **32** 510

IOP Publishing

Charlie Ironside, Bruno Romeira and José Figueiredo

Chapter 3

Resonant tunnelling diode: electrical and optical properties

Resonant tunnelling diodes (RTDs) are semiconductor devices that are made from epitaxial layers of semiconductor alloys with critical dimensions of less than 10 nm. This chapter covers how this epilayer construction affects some of the key electrical and optical properties that are employed in RTD photonic devices. Quantum tunnelling gives rise to negative differential conductance (NDC) operation in sub-picosecond time scales, so that RTDs can be used as the active component in electronic oscillators operating up to 1.93 THz. The optical properties of the epilayers reflect the usual optical properties of heterostructured semiconductors and RTD photonic devices can be employed in the detection, generation and modulation of light.

3.1 Introduction

In this chapter, we cover the electrical and optical properties of RTD-based structures relevant to understanding the photonic applications of these devices. As described in the previous chapter, RTDs are semiconductor devices, and their optical and electrical properties are derived from their semiconductor nature. We start this chapter on the assumption that the reader has a basic understanding of these properties, as can be obtained from textbooks on semiconductor optoelectronics, see for example [1].

3.2 Differential negative conductance

NDC can be observed by measuring the nonlinear current–voltage (I–V) characteristic of an RTD, as shown in figure 3.1. This corresponds to the key electrical characteristic employed in RTD photonic devices.

When direct current (DC) biased to an appropriate voltage, an NDC device exhibits negative conductance, i.e., the current decreases as the voltage is increased. In the NDC

doi:10.1088/978-0-7503-5714-2ch3

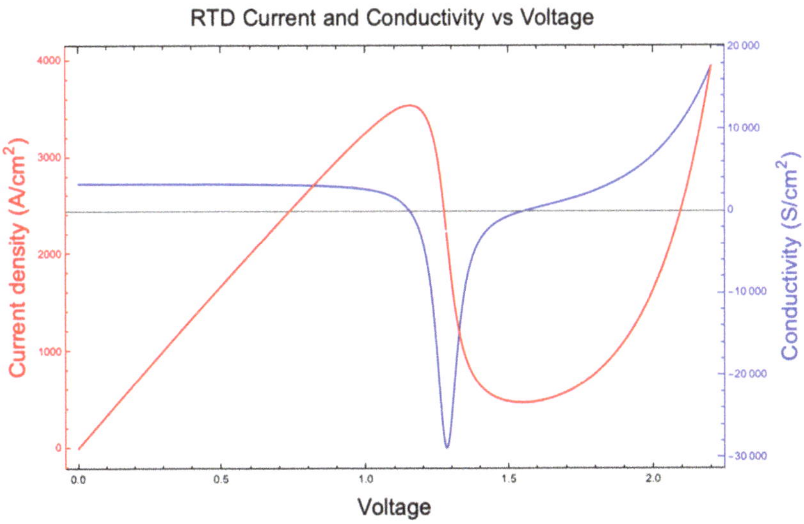

Figure 3.1. Red: The fit to the current–voltage (*I–V*) curve of an InGas/AlAs RTD (taken from [2]). Blue: The differential (with respect to V) of the *I–V* curve to give the conductivity—the NDC region is evident for bias voltages >1.2 V.

region of the *I–V* curve, the device has electrical gain, i.e., it generates alternating current (AC) power rather than absorbs power. For example, in figure 3.1 when biased around 1.1 V, the device has electrical gain. In the case of electrical feedback, via (for example) parasitic components (e.g. inductor line) and in the situation where the gain of the system overcomes the losses, the RTD will oscillate. The RTD can also be embedded in electric circuits, where oscillation is completely suppressed at all frequencies, but the circuit still provides linear or nonlinear electrical gain.

One of the really exciting aspects of RTDs is that electronic oscillators based on RTDs have operated at up to 1.93 THz. This frequency is the highest frequency reported for a room-temperature electronic single oscillator to date [3]. Therefore, RTDs exhibit the largest gain bandwidth of any room-temperature electronic device.

The oscillation frequency is determined by a mixture of the physical mechanisms underlying the NDC and the nature of the electrical feedback. In the RTD, the speed of the quantum tunnelling mechanism underlying the NDC is fundamentally limited by the dwell time or resonance time associated with the resonance state (see chapter 2). The dwell time is the average time the electron is confined by the barriers in the quantum well and is typically less than 1 ps. From the oscillation result, this time can be less than ~0.5 ps, and the device has a large intrinsic gain bandwidth. Consequently, RTD oscillators can exhibit the highest oscillation frequencies. The relative contributions of parasitic and fundamental contributions to the limit of the gain bandwidth can be challenging to disentangle, but it is currently the view [3] that the oscillation of RTD oscillators can be extended above 2 THz, securing their position as the fastest electronic device operating at room temperature.

The origins of the RTD *I–V* curve and the NDC lie entirely in the quantum effect of tunnelling through two energy barriers separated by a layer with a thickness less than the electron phase breaking distance. Therefore, to enable the quantum tunnelling effect, the dimensions of the epitaxial layers that make up the RTD must be small enough to keep the electron wavefunction coherent, as described in chapter 1 in the case of mesoscopic devices. An example of a layer structure of an RTD in the InGaAs/AlAs material system is illustrated in figure 3.2.

The explanation of how the RTD structure gives rise to the nonlinear *I–V* curve in figure 3.1 starts with figure 3.3, which shows the conduction band profile of a generic double barrier RTD with zero applied voltage.

As illustrated in figure 3.4, when a positive emitter voltage is applied to the RTD, the bottom of the conduction band in the emitter region is raised in energy relative to the collector side until it reaches the same energy as the resonance level, at the voltage level, V_p, this is the resonance condition and the RTD reaches the peak current flow, I_p. When the voltage is increased beyond, $V > V_p$, the RTD is in an off-resonance state and the current drops, reaching its lowest value I_v at the voltage V_v.

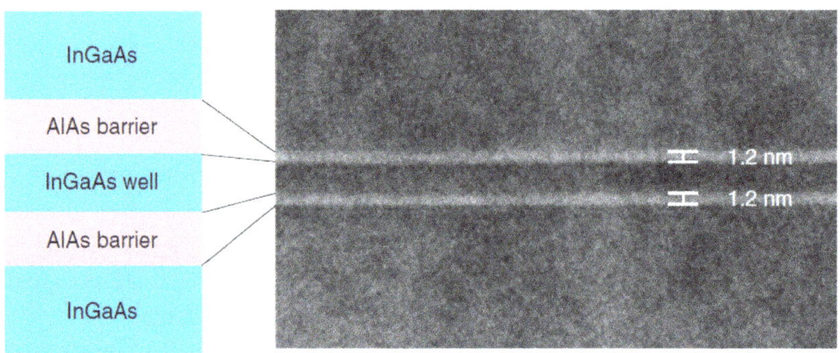

Figure 3.2. Adapted from [4]. Schematic of an InGaAs/AlAs double barrier RTD (left) and the cross-sectional transmission electron micrograph (TEM) image of the double barrier RTD (right).

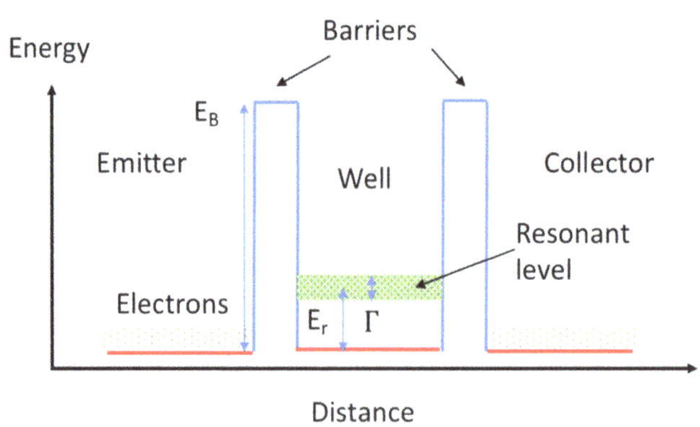

Figure 3.3. Schematic of the conduction band profile of the generic double barrier RTD with no voltage applied. The bottom of the conduction is plotted as a function of distance. The low bandgap material (red; e.g. InGaAs) forms a quantum well between two barriers of high bandgap material (blue; e.g. AlAs), Γ is the width of the electronic bound state resonance at energy E_r.

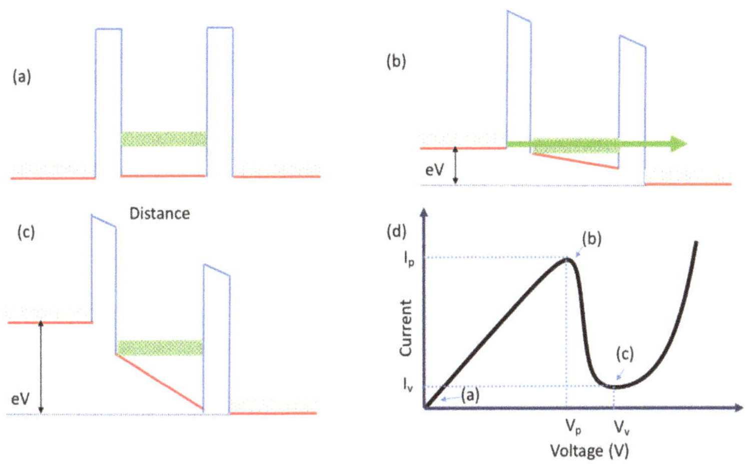

Figure 3.4. Conduction band profiles at three different bias voltages: (a) zero applied voltage $V = 0$, (b) resonance ($V = V_p$), (c) off resonance ($V = V_V$) and (d) schematic of the current–voltage curve of the RTD showing how the curve relates to the conduction band profiles.

The physics that explains the RTD current voltage I–V in detail starts by considering the various quantum mechanisms or channels open to the electrons as they make their way through the RTD structure. There has been much theoretical interest in the RTD and a good review can be found in [5]. Here we adopt a simplified explanation that is sufficient to describe the broad features of the RTDs analysed here and, with the help of some adjustable parameters, provides a reasonably good fit to the experimental I–V curve, based on references [2, 6].

A general principle of quantum transport in mesoscopic systems is the Landauer formula [7], $G = G_0 \sum_i^{All} T_i(E, V)$, where G is the electrical conductance and G_0 is a constant, and $T_i(E, V)$ a channel transmission coefficient. This formula relates the electrical conductance of a quantum conductor to the sum of the transmission coefficients for the various transmission channels. In what follows, we divide the transmission channels into just two channels: a resonant channel and a nonresonant channel. The resonant channel accounts for transmission through the resonant state in the quantum well. All of the other possible channels, e.g., transmission by high energy electrons over the barriers, are lumped into nonresonant channels. It turns out that this treatment is sufficient to account for the main features of the I–V curve and, with the aid of fitting parameters, it can be adapted to fit the measured I–V curves.

For transport through the resonance level in the quantum well, the standard formula for RTD the current density in the effective mass approximation is given by [2]

$$J_R = \frac{|e|m^*}{2\pi^2\hbar^3} \int_0^\infty T(E, V) \ln\left[\frac{1 + e^{(E_F - E)/kT}}{1 + e^{(E_F - E - eV)/kT}}\right] dE \tag{3.1}$$

where $|e|$ is the absolute value of the charge of an electron, m^* is the effective mass of the electron in the semiconductor, $\hbar$ is Planck's constant divided by 2π, E is the energy measured from the conduction band edge, and E_F is the Fermi energy. $T(E, V)$ is given by:

$$T_r(E, V) = \frac{\left(\frac{\Gamma}{2}\right)^2}{\left[E - \left(E_r - \frac{eV}{2}\right)\right]^2 + \left(\frac{\Gamma}{2}\right)^2} \tag{3.2}$$

where Γ is the energy resonance width and E_r is the energy of the resonant level relative to the bottom of the quantum well.

Generally, the width of the quantum well resonance, Γ, is small compared to room-temperature thermal energies and so $T_r(E, V)$ is negligible except when $E = \left(E_r - \frac{eV}{2}\right)$ applies. So, we can substitute $\left(E_r - \frac{eV}{2}\right)$ for E and take it outside the integral. The integral is then solved to get:

$$J_R(V) = \frac{|e| m \times T\Gamma}{4\pi^2\hbar^3} \ln\left[\frac{1 + e^{(E_F - E + eV)/kT}}{1 + e^{(E_F - E - eV)/kT}}\right] \cdot \left[\frac{\pi}{2} + \tan^{-1}\left(\frac{E_r - \frac{eV}{2}}{\frac{\Gamma}{2}}\right)\right] \tag{3.3}$$

So, this expression of the tunnelling current gives a suitable form of the equation that can be related to the physics. To model a given I–V curve, adjustable parameters are introduced as follows:

$$J_r(V) = A \ln\left[\frac{1 + e^{(B - C + n_1 V)e/kT}}{1 + e^{(E_F - E - n_1 V)e/kT}}\right] \cdot \left[\frac{\pi}{2} + \tan^{-1}\left(\frac{C - n_1 V}{D}\right)\right] \tag{3.4}$$

where A is in A cm^{-2}, B and C and are in V, and n_1 is dimensionless. Other nonresonant channels allow the electron to make its way through the double barrier RTD structure, such as thermal excitation over a barrier, and these give rise to a valley current that can be taken into account using the following expression:

$$J_{NR}(V) = H(e^{(n_2 V)e/kT} - 1) \tag{3.5}$$

where H is in A cm^{-2} and n_2 is dimensionless.

The total current density flowing through the RTD is given by $J(V) = J_R(V) + J_{NR}(V)$. Figure 3.5 shows the fit from [2] for a RTD made from AlGaAs/GaAs/AlGaAs (panel a) and AlAs/InGaAlAs/AlAs (panel b) semiconductor materials. Although this method allows any RTD I–V curve to be fitted, the fitting parameters obscure some of the underlying physics and only really hint at how the RTD design can be adjusted for any given application.

The final form of RTD current-voltage curve [8] is given by:

$$F(V) = I_R(V) + I_{NR}(V) = M[J_R(V) + J_{NR}(V)] \tag{3.6}$$

where M is a scaling factor that takes into account the area of the RTD device.

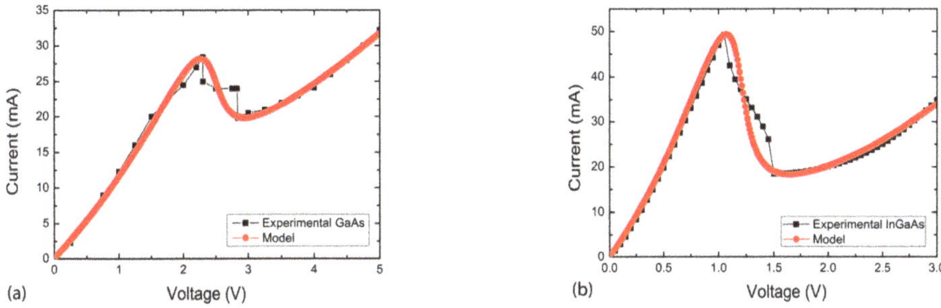

Figure 3.5. The current-voltage curves with a model fit of (a) AlGaAs/GaAs/AlGaAs RTD and (b) AlAs/InGaAlAs/AlAs RTD. The red line fit is from model summarised in equation (3.6). For (a) $M = 2 \times 10^{-6}$ cm^2, A = 45 000 A cm^{-2}, B = 0.05, C = 0.0874, D = 0.007, n_1 = 0.0350, H = 18 500 A cm^{-2}, and (b) $M = 1 \times 10^{-6}$ cm^2, A = 200 000 A cm^{-2}, B = 0.068, C = 0.1035, D = 0.009, n_1 = 0.0350, H = 8500 A cm^{-2}.

For many applications, a large peak I_p to valley current I_v ratio, $\left(\frac{I_P}{I_V}\right)$, is required, which can be obtained with large energy barriers height, E_B. If we compare figure 3.5(a) the I–V curve of the AlGaAs/GaAs/AlGaAs RTD with figure 3.5(b) the I–V curve of the AlAs/InGaAlAs/AlAs RTD, then it is apparent that the AlAs/InGaAlAs/AlAs RTD has a larger $\left(\frac{I_P}{I_V}\right)$, which can be attributed to the larger energy barriers height, E_B, and the smaller electron effective mass of the AlAs/InGaAlAs/AlAs RTD.

Figure 3.5 shows regions in the I–V curve where the device oscillates. In the I–V curve, the oscillation regions are where there are plateaus in the I–V curve at voltages larger than V_p. In this region of the I–V curve, the RTD has electronic gain. In addition, the parasitic circuit elements, associated with the DC bias circuit, provide feedback. The RTD has a large bandwidth (see the discussion in the next section), and therefore there is often a frequency with sufficient feedback and gain to cause the RTD to oscillate. The methods of obtaining an I–V curve of a tunnel diode under stable non-oscillatory conditions are discussed in [9], and can also be applied to RTD devices. These methods are, in essence, choosing the correct value of a series resistance in the RTD DC bias circuit.

The corresponding AC circuit of an RTD oscillatory circuit is depicted in figure 3.6. The time differential of the voltage $\dot{V}$ is given by:

$$\dot{V} = \frac{1}{C}[I - F(V)] \tag{3.7}$$

And the current is given by

$$\dot{I} = \frac{1}{L}[V_{DC} - RI - V] \tag{3.8}$$

After some algebraic manipulation of the above equations, the following second order differential equation can be obtained:

$$\ddot{V} + H(V)\dot{V} + G(V) = 0 \tag{3.9}$$

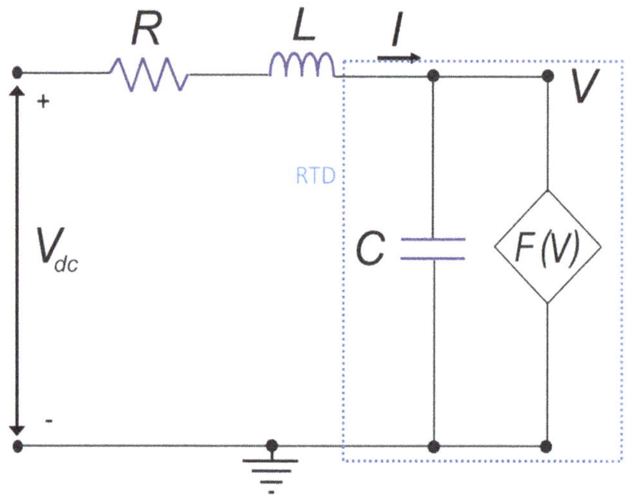

Figure 3.6. The DC bias circuit for the RTD, the *R* and *L* lumped components represent the parasitic components associated with the leads supplying the DC voltage to the RTD—the equivalent circuit for the RTD (inside the dashed blue box) is a capacitance in parallel with a voltage dependent current source $F(V)$.

With $H(V) = \frac{R}{L} + \frac{1}{C}\frac{dF(V)}{dV}$ and $G(V) = V - V_{DC} + RF(V)$. Equation (3.9) is a nonlinear second order differential equation that can be used to describe a Liénard system [10]. The Liénard system is a more general case of the more familiar Van der Pol oscillator [11] system of equations (3.7)–(3.9). In the Van der Pol case, the term multiplying the first order differential, $\dot{V}$, is a third order polynomial and with the Liénard system this nonlinear term can be more general, e.g. a higher order polynomial. If we evoke the analogy of a mechanical oscillator, then $G(V)$ is a nonlinear force and $H(V)\dot{V}$ is a damping factor.

When biased to the NDC region, the circuit illustrated in figure 3.6 acts as a relaxation oscillator operating at a fundamental frequency close to $f_0 \sim 2\pi(L\,C)^{-\frac{1}{2}}$, the condition for operation is that the series resistance is less than the absolute value of the negative resistance $R < |\,NDR\,|$.

3.2.1 High-frequency response

A key driver of research interest in resonant tunnelling devices is the potentially ultrahigh frequency response of the device. Fundamentally, the high-frequency response [3] can be traced back to the tunnelling time because the maximum frequency response cannot exceed the inverse of the tunnelling time. The tunnelling time sets the upper speed limit [5]. Theoretically, estimations of the tunnelling time through an RTD structure are in the range of 10–100 fs [12] and so, because the tunnelling time is so short, in any practical RTD circuit, the actual speed limit is de facto set by the parasitic components associated with the bias and feedback circuit.

Experimental investigation of the switching speed is consistent with the picosecond time scale for RTD switching [13], and it has proved possible to separately determine the contribution of parasitic or resistance/capacitance (*RC*) and the more fundamental contribution to the switching time from the quantum well quasibound state lifetime [14].

A simple intuitive approach to estimating the tunnelling time is to estimate the time that the tunnelling electron takes across the width of the barrier. This is calculated from the velocity of the electron, as determined from its k vector and the width of the barrier. Thus, the tunnel time is given as follows [15]:

$$t_{\text{tun}} = \left(\frac{2E}{m^*}\right)^{-1/2} W \tag{3.10}$$

where W is the barrier width and E is the electron energy. This was one of the original approaches to estimating the tunnel time through a single barrier.

To illustrate these considerations, we take as an example a high-frequency design of an RTD device with 1.9 THz oscillation frequency and ~0.4 μW output power [3]. The design principles can be summarised as follows:

- Reduce the thickness of the RTD layers: the AlAs energy barriers are 1 nm thick and the InGaAs well is 2.5 nm.
- Reduce the parasitic component values: the inductive and resistive load of the slot antenna for the THz radiation is minimised by using a microfabricated slot antenna and using highly doped contact layers (InGaAs n-type $\sim 5 \times 10^{19}$ cm^{-3}).
- For high power output at high frequency, a large gain bandwidth and then a large peak to valley ratio is required. The high AlAs energy barriers ensure a deep quantum well that reduces the valley current by lowering the thermal energy-induced leakage of charge carriers over the barriers.

3.3 Optical properties: optical waveguiding, electroabsorption, photoconduction lasers and LEDs

The optical properties of the RTD photonic devices are dominated by the optical properties of the epitaxial semiconductor alloy layers that make up the device. The actual RTD structure consists of layers that are much thinner than the wavelength of light in the semiconductor. For example, a typical RTD structure, including barriers and quantum well, is <10 nm thick and RTD photonic devices designed for optical telecommunications wavelengths (~1500 nm) have an optical wavelength of around 400 nm in the device. So, the RTD structure is only a minor perturbation of the optical properties. The RTD is mainly included to provide very high-speed switching and/or high bandwidth NDR. RTD photonic devices have been fabricated with optical functionalities that include optical waveguide, electroabsorption, photoconduction and lasers—the device optical functionality is provided by the other semiconductor alloy epilayers that make up the device. For example, if an RTD electroabsorption optical modulator is required, then the device needs semiconductor epilayers that provide optical waveguiding so that the light is confined to a core layer for the optimum interaction length and the bandgap energy of the core layer is tuned to the value appropriate for maximum electroabsorption effect.

In this chapter, we give a general description of the optical properties that are useful for an RTD photonics device. In subsequent chapters, more specific descriptions are provided of exactly how these properties are employed in the following devices:

(i) Optical waveguide RTD electroabsorption modulator (RTD-EAM),
(ii) RTD photodetector (RTD-PD),
(iii) RTD laser diode (RTD-LED),
(iv) RTD light-emitting diode (RTD-LED).

3.3.1 Optical waveguiding

Semiconductor electroabsorption modulators [16] can be employed in optical communication systems and are typically optical waveguide devices because in waveguide devices the interaction length between the core semiconductor epilayer and the light is not limited by diffraction, as would be the case in bulk optics. Diffraction limits the interaction region in bulk optics because the focal region is inversely proportional to the width of the focus, so a tightly focussed beam rapidly expands. In a waveguide configuration, the focal region is captured by the waveguide and tight focus is maintained for the length of the waveguide. The waveguide device is limited not by diffraction but by losses in the waveguide, be they scattering or absorption losses. The RTD-EAM described in chapter 4 is a waveguide and here we describe the basic principles of optical waveguiding for semiconductor photonic devices.

Figure 3.7 shows a schematic of a generic semiconductor optical waveguide. The propagation modes for light in waveguides can be calculated (see figure 3.8) on the Simulations in Integrated Optics website [17].

The critical advantage conferred by the waveguide configuration is that the interaction length between light and the semiconductor material in the core is not limited by diffraction but by the optical losses that can be scattered from (for example)

Figure 3.7. Layout and epilayer structure of a generic RTD semiconductor optical waveguide. For the intensity distribution of the fundamental mode, see figure 3.8.

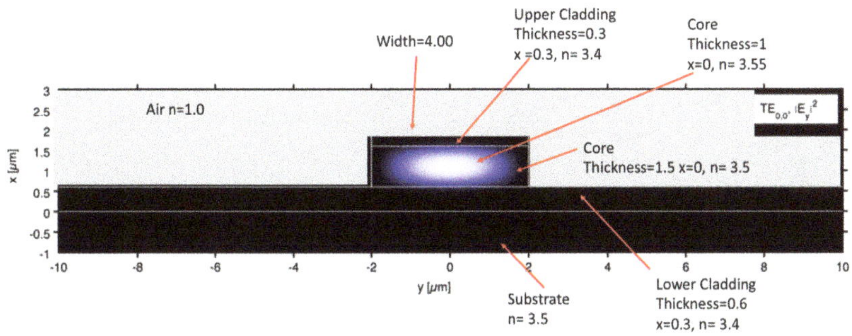

Figure 3.8. Cross-sectional view of a Al_xGa_{1-x} As waveguide. The intensity distribution for the fundamental mode (TE00) for a wavelength of 900 nm is shown in blue. Dimensions are given in microns. The upper and lower cladding are $Al_{0.3}Ga_{0.7}As$ (n = 3.4), and the core is GaAs (n = 3.55). The mode is calculated using the EIMS section of [17]. For the refractive index of $Al_xGa_{1-x}As$ as a function of alloy fraction and wavelength plus guidance on how to use the EIMS section of [17] for $Al_xGa_{1-x}As$ optical waveguides, see chapter 11 in section 11.1.3 the corresponding file: Waveguide Modes.ipnb.

rough walls on the channel mesa or from intrinsic losses from absorption by the semiconductor. The increased interaction length is essential for electroabsorption devices and can be crucial in photodetector devices. At photon energies just below the bandgap energy of the semiconductor, which are dominated by scattering losses, the semiconductor waveguide loss has been measured to be around 20 cm^{-1} [18].

Another advantage of the waveguide configuration is the compact layout of the electrodes in the dimension perpendicular to the plane of the layers. So, for example, a small applied voltage of a few volts can result in electric fields of a few kV cm^{-1} in the core of the waveguide, which is useful for both electroabsorption and electro-optic devices.

3.3.2 Electroabsorption

The RTD can efficiently switch the electric field in the core part of a semiconductor optical waveguide—the exact mechanism is discussed in chapter 4. The change in an electric field can be used to alter the absorption of light in the core of the semiconductor optical waveguide, and thereby encode the light with information. The nonlinear electrical switching characteristics of the RTD, as illustrated in figure 3.1, mean that the RTD electric field switching is best for digital modulation rather than analogue optical modulation.

The Franz–Keldysh effect refers to electrical field dependent change in the optical absorption close to the band edge of a bulk semiconductor. From a consideration of how an applied electrical field induces overlap between the electron valence band and conduction band wavefunctions in a material [19–21], the following expression for the absorption coefficient spectrum can be derived

$$\alpha(\omega, F) = \left(\frac{C_p \theta_F^{\frac{1}{2}}}{\omega}\right)\left[\,|\,\acute{A}i(\eta)|^2 - \eta\,|\,Ai(\eta)\,|^2\right] \qquad (3.11)$$

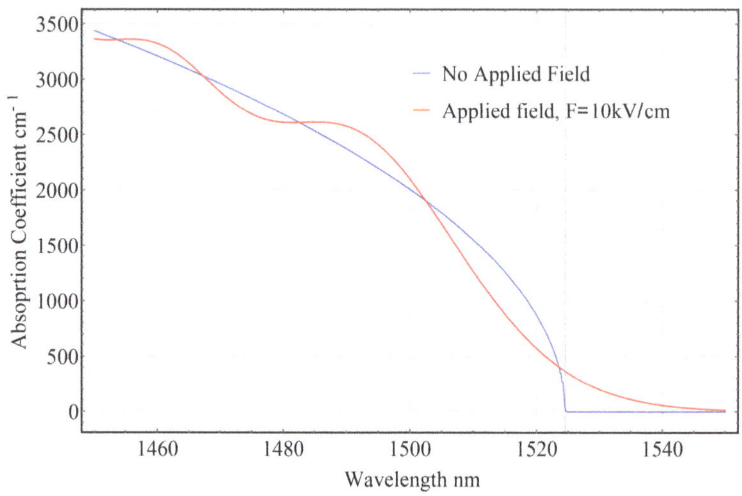

Figure 3.9. Franz–Keldysh effect in InAlGaAs. Calculated optical absorption spectrum of a semiconductor alloy $In_{1-x-y}Ga_xAl_yAs$ [22] $y = 0.05$ with and without an applied electric field.

where $\theta_F^3 = \frac{e^2F^2}{2m_{vc}\hbar}$, $\eta = (E_g - \hbar\omega)/\hbar\theta_F$, F is the applied electric field, C_p is a material constant that is similar for III–V semiconductors and CGS units is given by $C \sim 1.76 \times 10^{12}$ cm^{-1} s$^{-1/2}$—actually, this value is for GaAs. The parameter m_{vc} is the reduced mass of the electron-hole pair for GaAs, $m_{vc} = 0.065\ m_o$, where m_o is the mass of the free electron; ω is the optical angular frequency and E_g is the bandgap energy; and, $Ai(\eta)$ is the Airy function and $\acute{A}i(\eta)$ is the differential of the Airy function (Airy Prime function).

The optical spectrum calculated from equation (3.11) is shown in figure 3.9 for the alloy $In_{1-x-y}Ga_xAl_yAs$ with the $y = 0.05$ fraction set so that the bandgap energy is close to the optical telecommunications 'C band' photon energies (wavelength around 1550 nm) [22].

It turns out that the Franz–Keldysh effect of bulk semiconductors is not the optimum electroabsorption effect for electroabsorption modulators. It is better to employ multiple quantum wells in a layered stack, in which case the electro-absorption effect is termed the quantum confined stark effect (QCSE) [23]. The best-performing EAMs are produced when the quantum wells are optimised to enhance the QCSE for a given wavelength [23, 24].

3.3.3 Photoconductivity and photodetection

The *I–V* curve of the RTD is, to some extent, dependent on the absorption of light in the device because the conductivity of the epilayers is increased by creating electron-hole pairs, as in the photoconductivity effect [25]. RTD devices can utilise photo-conductivity to produce highly sensitive photodetectors [26] in the visible and near-infrared. This concept has recently been extended to include the mid-infrared [27].

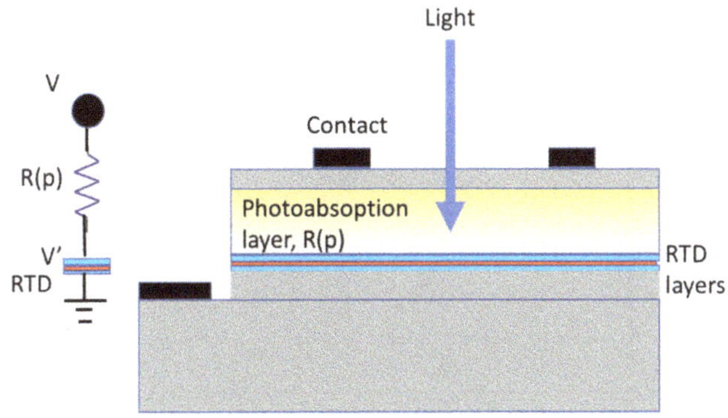

Figure 3.10. The lumped element electrical circuit and the layer structure of a generic RTD photoconductive circuit. A voltage, *V*, is applied to bias the circuit and voltage, *V'*, which depends on the light intensity modulated resistance, *R*(*p*), appears at the double barrier RTD structure.

We will give a simple analysis of how a double barrier RTD structure can amplify the current in a photoconductive device.

Photons with energy greater than the bandgap energy create mobile charge carriers in a layer of the RTD device (note that much lower carriers are generated in the RTD barriers and quantum well, which are generally too thin to absorb a significant amount of light), and thereby increase the electrical conductivity of the photoabsorption layer. A simple way to analyse this [28] is to include a light dependent resistance component in the circuit, as shown in figure 3.10. The total resistance of the RTD device internal resistance now becomes:

$$R_s = R_i + R(p) \tag{3.12}$$

where the total resistance has been split into a light dependent part, $R(p)$, and a part independent of the incident light R_i.

The voltage, V', across the double barrier structure is given by:

$$V' = V - R(p)_s I \tag{3.13}$$

where I is the current in the device.

When light is incident on the device, the effective voltage, V', at the double barrier is increased as the resistance, $R(p)_s$, is reduced by photoconductivity. If the device is biased close to resonance then, with a slight change in incident light, the device can jump from on to off resonance, i.e., from V_p to V_v (see figure 3.4), thus providing amplification. Therefore, the effective responsivity of the RTD photodetector (RTD-PD) increases when the device is biased close to V_p. This is illustrated in figure 3.11, where it is shown that the responsivity as a function of the applied voltage can be enhanced by a factor ~10 when the device is biased in the valley region as compared to an applied voltage of ~0.5 V.

Subsequently, more refinements [29] to the RTD-PD have achieved responsivities as large as $3.1 \times 10^4\ W^{-1}$ and the ability to detect a few (~26) photons. The responsivity is large enough for the RTD-PD to directly drive other optoelectronic

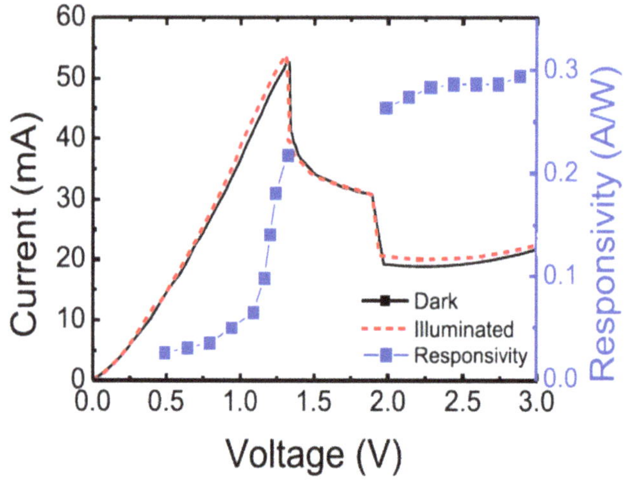

Figure 3.11. Adapted from [26], this figure shows the dark and illuminated *I–V* curve for a waveguide RTD-PD with an InGaAlAs photoabsorption layer and a AlAs/InAlAs/AlAs RTD. The effective responsivity is shown versus applied bias voltage for light with a wavelength of 1550 nm.

devices, such as laser diodes [30], and thus build systems such as the optoelectronic oscillator [31].

There are other performance properties apart from responsivity, which will be covered in more detail in chapter 5 on the RTD-PD.

3.3.4 Optical emission

The RTD can also be incorporated into semiconductor devices that emit light. RTDs have been incorporated into laser diodes (RTD-LD) [30] and light emitting diodes (LEDs)—RTD-LEDs [32]. Usually, the light emission requires that the RTD is combined with a *pn* junction but it has also recently been observed in devices without *p*-type material and operates via impact ionisation [33].

The RTD confers an NDC region to the electrical characteristics of the RTD-LD and RTD-LED. The NDC region can be used to amplify an applied modulation signal so that the optical output from the device can be modulated with a lower power and higher frequency signal. For the RTD-LED, this can lead to a modulation figure of merit, M (in μA/mA × MHz), increasing from $M{\sim}6$ in heavily doped LEDs to $M{\sim}4400$ in RTD-LEDs [34].

In chapter 6, we will extensively discuss how the RTD-LD improves the modulation of laser diodes, and on how electrons in the high-field DBQW structure are promoted to the conduction band via an impact ionisation mechanism thus creating electron–hole pairs that can recombine via spontaneous emission of light.

3.4 Conclusions

This chapter has introduced the essential electrical and optical characteristics of semiconductor optoelectronic devices, including a double barrier quantum well

RTD structure. The unifying theme is that the RTD provides very high bandwidth electrical gain in a compact epilayer structure (~10 nm) that is easily incorporated with other semiconductor epilayers. This enables unique photonic device functionalities in photodiodes, lasers, LEDs, and modulators that can be driven with low amplitude and high-speed input signals, taking advantage of the wide bandwidth and on-chip electrical gain of RTD-integrated devices.

References

[1] Bhattacharya P 1996 *Semiconductor Optoelectronic Devices* 2nd edn (Pearson)

[2] Schulman J N, Santos H J D and Chow D H 1996 Physics-based RTD current–voltage equation *IEEE Electron Device Lett.* **17** 220–2

[3] Maekawa T, Kanaya H, Suzuki S and Asada M 2016 Oscillation up to 1.92 THz in resonant tunneling diode by reduced conduction loss *Appl. Phys. Express* **9** 024101

[4] Sugiyama H, Suzuki S and Asada M 2011 Room-temperature resonant-tunneling-diode terahertz oscillator based on precisely controlled semiconductor epitaxial growth technology *NTT Tech. Rev.* **9** 1–7 https://www.ntt-review.jp/archive/ntttechnical.php?contents=ntr201110fa2.html

[5] Bandyopadhyay S 2012 *Physics of Nanostructured Solid State Devices* (Berlin: Springer)

[6] Sun J P, Haddad G I, Mazumder P and Schulman J N 1998 Resonant tunneling diodes: models and properties *Proc. IEEE* **86** 641–61

[7] Blanter Y M and Nazarov Y V (ed) 2009 *Quantum Transport: Introduction to Nanoscience* (Cambridge: Cambridge University Press) pp 7–123

[8] Figueiredo J, Romeira B, Slight T and Ironside C 2010 *Resonant Tunnelling Optoelectronic Circuits* (Advances in Optical and Photonic Devices) (InTech Open)

[9] Wang L Q, Figueiredo J M L, Ironside C N and Wasige E 2011 DC characterization of tunnel diodes under stable non-oscillatory circuit conditions *IEEE Trans. Electron Devices* **58** 343–7

[10] Slight T J, Romeira B, Wang L Q, Figueiredo J M L, Wasige E and Ironside C N 2008 A lienard oscillator resonant tunnelling diode-laser diode hybrid integrated circuit: model and experiment *IEEE J. Quantum Electron.* **44** 1158–63

[11] Guckenheimer J 1980 Dynamics of the Van-Der-Pol equation *IEEE Trans. Circuits Syst.* **27** 983–9

[12] Barker J R 1985 Quantum theory of hot-electron tunnelling in microstructures *Physica B+C* **134** 22–31

[13] Whitaker J F, Mourou G A, Sollner T and Goodhue W D 1988 Picosecond switching time measurement of a resonant tunneling diode *Appl. Phys. Lett.* **53** 385–7

[14] Growden T A, Brown E R, Zhang W D, Droopad R and Berger P R 2015 Experimental determination of quantum-well lifetime effect on large-signal resonant tunneling diode switching time *Appl. Phys. Lett.* **107** 153506

[15] MacColl L A 1932 Note on the transmission and reflection of wave packets by potential barriers *Phys. Rev.* **40** 621–6

[16] Shimomura K and Arai S 1994 Semiconductor wave-guide optical switches and modulators *Fiber Integr. Opt.* **13** 65–100

[17] Hammer M 2023 *Simulations in Integrated Optics* (https://www.computational-photonics.eu/) (accessed 2023)

[18] Jain M and Ironside C N 2003 Internal optical loss measurements in InGaAs-InAlGaAs quantum-well lasers operating around 1550 nm *IEEE Photonics Technol. Lett.* **15** 631–3

[19] Tharmalingam K 1963 Optical absorption in the presence of a uniform field *Phys. Rev.* **130** 2204–6
[20] Seraphin B O and Bottka N 1965 Franz–Keldysh effect of the refractive index in semiconductors *Phys. Rev.* **139** A560–5
[21] Ironside C 2017 *Semiconductor Integrated Optics for Switching Light* (Morgan & Claypool)
[22] Figueiredo J M L 2000 Optoelectronic properties of resonant tunelling diodes *PhD Thesis* (Porto, Portugal: Universidade do Porto)
[23] Chin M K 1995 Comparative-analysis of the performance limits of Franz–Keldysh effect and quantum-confined stark-effect electroabsorption wave-guide modulators *IEE Proc., Optoelectron.* **142** 109–14
[24] Jain M and Ironside C N 2007 A multi-section technique for the electroabsorption measurements in waveguide semiconductor electroabsorption modulators *IET Optelectron.* **1** 163–8
[25] Moss T S 1965 Photoconductivity *Rep. Prog. Phys.* **28** 15
[26] Romeira B, Pessoa L M, Salgado H M, Ironside C N and Figueiredo J M L 2013 Photodetectors integrated with resonant tunneling diodes *Sensors* **13** 9464–82
[27] Rothmayr F *et al* 2018 Mid-infrared GaSb-based resonant tunneling diode photodetectors for gas sensing applications *Appl. Phys. Lett.* **112** 161107
[28] Coelho I J S, Martins J F, Figueiredo J M L and Ironside C N 2004 Modeling of light-sensitive resonant-tunneling-diode devices *J. Appl. Phys.* **95** 8258–63
[29] Pfenning A, Hartmann F, Langer F, Hofling S, Kamp M and Worschech L 10 2014 Cavity-enhanced resonant tunneling photodetector at telecommunication wavelengths *Appl. Phys. Lett.* **104** 101109
[30] Slight T J and Ironside C N 2007 Investigation into the integration of a resonant tunnelling diode and an optical communications laser: model and experiment *IEEE J. Quantum Electron.* **43** 580–7
[31] Romeira B, Seunarine K, Ironside C N, Kelly A E and Figueiredo J M L 2011 A self-synchronised optoelectronic oscillator based on an RTD photodetector and a laser diode *IEEE Photon. Technol. Lett.* **23** 1148–50
[32] Vanhoof C, Genoe J, Mertens R, Borghs G and Goovaerts E 1992 Electroluminescence from bipolar resonant tunneling diodes *Appl. Phys. Lett.* **60** 77–9
[33] Hartmann F *et al* 2017 Temperature tuning from direct to inverted bistable electroluminescence in resonant tunneling diodes *J. Appl. Phys.* **122** 154502
[34] VanHoof C, DeNeve H, Mertens R, Romandic I, Goovaerts E and Borghs G 1997 Gigahertz modulation of tunneling-based GaAs light emitters *IEEE Photon. Technol. Lett.* **9** 1463–5

IOP Publishing

Charlie Ironside, Bruno Romeira and José Figueiredo

Chapter 4

Resonant tunnelling diode-electroabsorption modulator (RTD-EAM)

Optical communication systems relying on microwave or millimetre-wave-optical transducers are fundamental to the development of applications such as low-cost fibre-optic communication networks, radio-over-fibre wireless communications and radio phased array antennas. These systems require very fast optical modulators [1] because laser diodes cannot be directly modulated at frequencies beyond their resonance frequency, which is typically below a few tens of gigahertz [2]. This chapter reviews our work on low driving voltage high-speed and highly efficient electrical active electroabsorption modulators (EAMs). EAMs are based on the integration of a double barrier quantum well (DBQW) structure within AlGaAs/GaAs and InGaAlAs/InGaAs unipolar semiconductor optical waveguides for operation at the optical transmission windows corresponding to wavelengths of around 900, 1310 and 1550 nm, respectively. These novel devices are known as resonant tunnelling diode electroabsorption modulators (RTD-EAMs).

4.1 Introduction

Optical communication systems supporting very high data bit rates per channel comprise a light-wave carrier source, typically a laser diode, a transmission media, such as the air or an optical fibre (eventually utilising in-line optical amplification), and a light-wave receiver incorporating a photodetector, such as a photodiode (PD) or an avalanche photodiode (APD).

The process that allows a light-wave carrier to convey information is known as optical modulation, by which a light-wave property is intentionally modified according to a high-frequency electrical signal (modulating signal) that carries the data to be transmitted. A light-wave carrier can be modulated in amplitude (or intensity), frequency, polarisation and phase, depending on the application and the technologies

doi:10.1088/978-0-7503-5714-2ch4

employed. This modulated light wave is transmitted by a transparent medium, such as the air or an optical fibre cable.

Assuming that the bandwidth of the optical communication system is not limited by the receiver's capacity to convert the intentional changes of the light-wave property, such as the intensity, into an electrical signal that resembles the modulating signal, the speed of the optical modulation determines the data bit rate that the system is capable of handling.

Generically, there are two methods of modulating the laser diode light-wave signal: direct and external. In direct modulation, the driving current of the laser diode changes according to the modulating signal to generate an intensity-varying optical signal. As the driving current changes, the variation of electrical charge concentration in the laser diode induces a refractive-index modulation. This leads to optical-frequency chirping and results in the widening of the laser diode's linewidth, i.e., the width of the power spectral density of the emitted laser optical spectrum in terms of frequency or wavelength, typically characterised by the full width at half-maximum.

This linewidth widening is known as chirp. The chirp parameter is defined as the ratio of the changes in the real and the imaginary part of the refractive index caused by the laser driving current. Therefore, direct modulation causes linewidth broadening, which introduces distortion in the transmitted signal and limits system performance, particularly in long distance communications and at very high data rates. The maximum speed achievable by the direct modulation is determined by the carrier lifetimes of spontaneous and stimulated emission, along with the photon lifetime in the laser diode cavity. Typically, direct modulation is used to realise data bit rates well below 20 gigabits per second (Gbit s^{-1}) [2]. In practical terms, direct modulation techniques are inappropriate for data rates above 2.5 Gbit s^{-1}. External modulation schemes need to be considered to attain data bit rates beyond a few Gbit s^{-1}. External modulation is primarily used when chirp becomes detrimental, but there is no obligation to use it only for high data rates.

In external modulation schemes, a separate device, known as an optical modulator, is used to modify a given property of the continuous-wave optical signal produced by an unperturbed laser diode. The optical signal produced by the laser diode travels through the modulator medium, whose properties, e.g., the complex refractive index, can be intentionally modified by an external agent, e.g., an electrical field. When a fast-changing electric field induces changes at the same rate in the real and/or imaginary parts of the complex refractive index, this fast change leads to changes in at least one of the light wave's measurable properties, e.g., the phase, intensity, polarisation or direction of propagation.

External optical modulators are of two types: electro-optical phase modulators, which are used when the changes in the real part of the complex refractive index are employed; and EAMs, which are used when the external perturbation modulates the imaginary part of the complex refractive index.

4.2 Electro-optic and electroabsorption modulators

In the external modulation simplest version, the laser diode driving current is kept unchanged and an optical modulator circuit is employed to modulate the intensity of the laser diode continuous-wave optical output according to the electrical signal carrying the information. This information carrying electrical signal modifies the modulator refractive index or the absorption coefficient, which produces a time-varying intensity optical signal that incorporates the data to be transmitted. The circuitry of the external modulator can be integrated with the laser diode or it can be a standalone device.

External modulators can be implemented either in waveguide configuration or in bulk in transverse transmission or reflection configurations [3]. The waveguide types are the most appropriate for compact transmitter and photonic integrated circuit implementations, in which waveguides are used to guide the input and output optical signals between devices along the surface of the substrate.

4.2.1 Electro-refraction modulators

When variations of the external electric field ΔE induce linear changes in the real part of the refractive index, $\Delta n \propto \Delta E$ (linear electro-optic effect), known as electro-refraction, this leads to light-wave phase changes. In electro-refraction modulators based on the linear electro-optic effect, an applied voltage induces a proportional change of the material refractive index, which leads to a variation of the light-wave phase: the light-wave induced phase change $\Delta\phi$ is directly proportional to the magnitude of the applied voltage, V: $\Delta\phi(V) \propto V$. This behaviour can be used to implement interferometric structures whose behaviour depends on the applied voltage.

The linear electro-optic effect is used to implement phase or intensity modulators, e.g., in an interferometer Mach–Zehnder configuration, as represented in figure 4.1(a) [4, 5]. The Mach–Zehnder configuration is typically implemented using dielectric ferroelectric materials such as lithium niobate ($LiNbO_3$) because they have a sizeable electro-optic effect. This leads to low loss and marginally associated heating effects compared to semiconductor materials.

Modulators based on ferroelectric materials are typically several centimetres long because of the need for interferometer or directional coupler configurations. Consequently, monolithic integration with photonic integrated circuits is

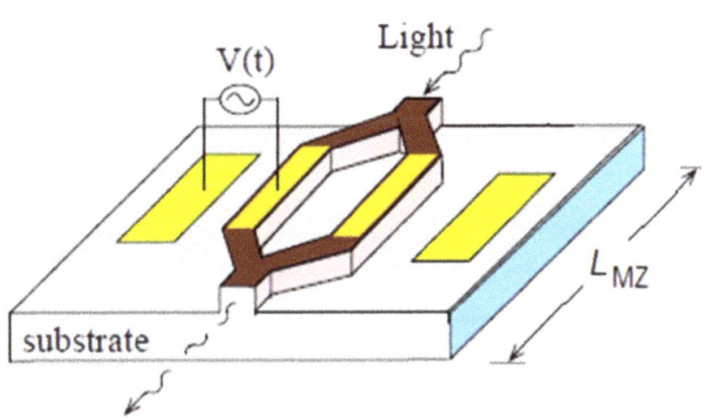

Figure 4.1. A typical external intensity modulator in a waveguide Mach–Zehnder interferometric configuration based on the linear electro-optic (electro-refraction) effect.

challenging. Additionally, they require driving peak-to-peak voltage amplitudes as large as 5 V at very high frequencies. Despite the limitations of compact solutions, bandwidths of more than 100 GHz can be achieved by employing travelling–wave designs in which the electrical signal is propagated on a micro-strip line along with the light wave [4, 5].

4.2.2 Electroabsorption modulators

When the electric field variations ΔE lead to changes in the imaginary part of the refractive index, and consequently on the material absorption coefficient α, the effect is known as electroabsorption. This produces a modulation of the transmission coefficient of the waveguide core material, enabling the modulation of the intensity of the light wave. In other words, the waveguide core transmission coefficient, T, dependent on the absorption coefficient α, is then modulated by a changing applied voltage ΔV: $T(\Delta V) \propto \exp[-\alpha(\Delta V) \cdot L]$, where L is the length of the waveguide region under the effect of the applied voltage, as indicated in figure 4.2, which shows schematically a section of the waveguide electroabsorption external modulator basic configuration.

The EAMs considered here are implemented using direct bandgap semiconductor materials. Typically, they are much smaller than interferometric electro-refraction modulators. Furthermore, because of their smaller size, compatibility with CMOS technologies and response times in the sub-picosecond, EAMs based on semiconductor waveguides are interesting components for photonic integrated circuit high-bandwidth optical communication links. Assuming that the parasitic of these components can be substantially reduced, the bandwidth of EAMs can reach hundreds of GHz.

Traditionally, semiconductor EAMs have been based on P–I–N semiconductor heterostructures. Semiconductor waveguide EAMs are designed to operate at optical wavelengths slightly below the bandgap of the low doped waveguide core or intrinsic material, making them transparent at these wavelengths. Under an

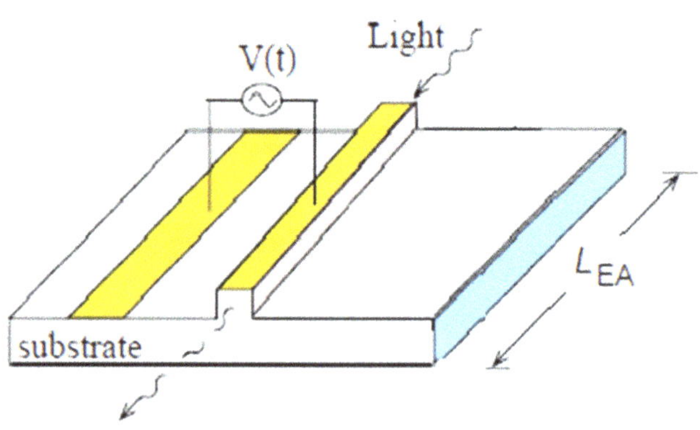

Figure 4.2. Typical configuration of an intensity modulator employing the electroabsorption effects in semiconductors (either Franz–Keldysh or quantum confined Stark effects). Typically, $L_{MZ} \gg L_{EA}$.

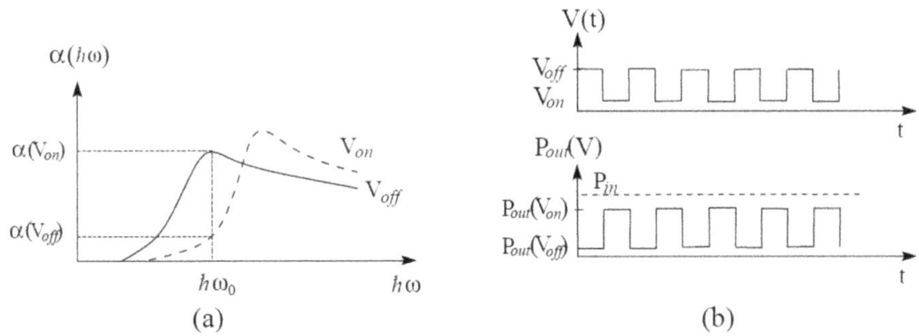

Figure 4.3. Schematic representation of the operation of an EAM: (a) absorption coefficient at two voltage values, showing the band-edge broadening due to the Franz–Keldysh or quantum confined Stark effects; (b) the applied voltage and the modulator output power. $\hbar\omega$ is the incident light photon energy; $V_{\rm on}$ and $V_{\rm off}$ represent the voltage values corresponding to the on and off states.

applied voltage, a band-edge shift is observed, resulting in an increase of the light-wave propagation loss, i.e., the applied voltage leads to a reduction of the optical intensity at the waveguide output. Figure 4.3 shows the operation principle of an electroabsorption modulator. In the on-state (low applied voltage), most of the propagating light reaches the waveguide output (high output intensity); while in the off-state (higher applied voltage), most of the light is absorbed as it propagates, leading to a low optical intensity at the waveguide output.

Although potentially with a small footprint, compatibility with CMOS technologies and enabling a response time in the sub-picosecond range, EAMs typically still require a large driving voltage. The applied voltage needed for an on/off ratio of about 10 dB is typically larger than 2 V (peak-to-peak voltage, $V_{\rm PP}$). A limiting factor in the high-speed response is the time constant for the holes to be swept out of the waveguide core.

4.3 Resonant tunnelling diode EAM device

Under operation, P–I–N semiconductor EAMs are voltage driven under reverse bias to induce the electric field change needed to modulate the absorption coefficient. As mentioned, a considerable voltage change at high frequency (gigahertz range) is required to achieve a significant intensity variation (around 2 $V_{\rm PP}$ to reach 10 dB intensity change).

As discussed previously, when a resonant tunnelling diode (RTD) operation point switches from the peak region to the valley region, a significant internal voltage drop enhancement results across the depletion region in the collector section adjacent to the DBQW, due to the current decrease from the peak to the valley (reduction through the NDC region), see figure 4.4(a). Consequently, assuming that there is no change in the series resistance, this voltage drop results in an electrical field enhancement, particularly across the collector section adjacent to the DBQW, which leads to an increment of the material absorption coefficient for light with photon energies close to but smaller than the collector's bandgap energy, through the Franz–Keldysh effect [6, 7], and then to an optical transmission reduction.

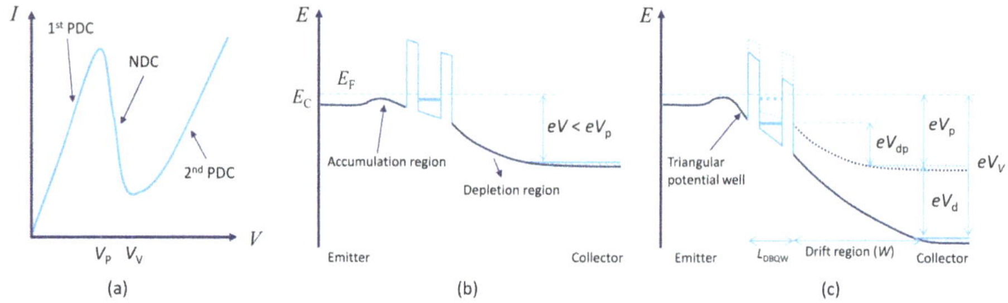

Figure 4.4. Effect of applied biased on RTD Γ-band: (a) *I–V* curve, (b) slightly before the peak and (c) on the valley.

However, in typical RTD implementations, this effect leads to negligible modulation depths because the thickness of the depleted region is at least one order of magnitude lower than the operating wavelength. Such a small light interaction volume means that a transverse transmission electroabsorption modulator, i.e., modulator where the light propagates perpendicularly to the epilayer plane, would produce a minimal absorption enhancement.

The concept of the RTD-EAM is based on a N–n–N heterostructure where the RTD sits in the low doped core region of the semiconductor optical waveguide and the light propagates along the epilayer plane, see figure 4.5. For an appropriate epilayer stack design, the overall RTD-EAM current–voltage characteristics resemble the baseline RTD *I–V* characteristic, see figure 4.4 (a).

The operation principle of an RTD-EAM can be understood as follows, with the help of the Franz–Keldysh effect [6, 7]. Basing the RTD-EAM slightly below the peak, the waveguide will be in a high transmission state. Applying a high-speed external voltage whose amplitude is sufficient to induce peak-to-valley transitions or biasing an RTD-EAM inserted in a resonant tank leads to the modulation of the inter-band absorption coefficient for light with photon energies close to but smaller than the collector's bandgap energy.

The electric field distribution dependence on the applied voltage in the direction of the carrier flow in an RTD structure can be understood considering the Γ-conduction band profile of the injection and collector spacer layers, as schematically shown in figure 4.4. Below resonance (first PDC region), the applied voltage is dropped mainly across the DBQW and its adjacent regions, particularly in the depletion zone formed on the collector side, see figure 4.4(b), and the electric field varies accordingly. Above resonance (in the NDC and the second PDC region), the additional applied bias voltage enhances the depletion of the collector region, reinforced by the current drop, which leads to a larger depletion region complemented by the electric field enhancement, see figure 4.4(c). In other words, most of the additional applied voltage beyond the peak voltage is dropped across this depletion region, where a strong electric field builds-in and/or extends across the collector's low doped region.

In the RTD-EAM transverse transmission configuration, the light would be injected perpendicularly to the tunnelling plane, either via the top or via the

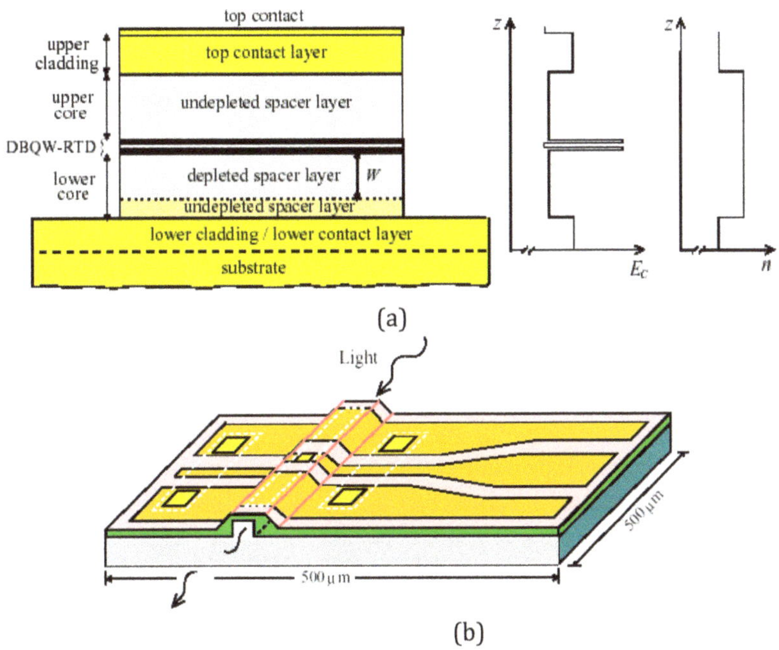

Figure 4.5. (a) Diagram of a unipolar resonant tunnelling diode optical waveguide wafer structure (RTD-OW), and the corresponding Γ-conduction band-edge and refractive-index profiles. (b) Ridged waveguide channel configuration. Such RTD-OWs, apart from the light confining layers (the lower refractive-index regions—top and bottom cladding layers), correspond to a DBQW-RTD with very thick undoped/low doped emitter and collector spacer layers.

substrate, see figure 4.5(a), which makes the light absorption volume relatively small, leading to a minimal modulation of the absorption coefficient. Thus, a small modulation depth is expected in the transmission. This limitation is easily overcome in a waveguide configuration, where the light propagates in the plane of the DBQW. The interaction volume can be additionally augmented by increasing the thickness of the undoped and low doped regions surrounding the DBQW. The light absorption volume is then substantially due to the increase of the light path in the direction parallel to the barrier-depletion region interface plane.

It is expected the RTD-EAM requires a much lower driving voltage for a given modulation depth than a P–I–N electroabsorption modulator by taking advantage of the electrical gain provided by the N-shape RTD-EAM current–voltage characteristic.

The combination of the fast-switching capabilities with the broadband electrical gain of RTD-based structures enables the implementation of high-speed, highly efficient and low-cost EAMs. RTD-EAMs based on AlGaAs/GaAs for operation at optical wavelengths around 900 nm and RTD-EAMs made of InGaAlAs/InP operating at 1550 nm have been demonstrated as highly efficient very low driving voltage EAMs [8, 9]. The modulation efficiency characterised by the bandwidth-to-drive-voltage ratio, for a 10 dB modulation depth, was higher than 40 GHz V^{-1}.

Moreover, the InGaAlAs/InP structures can be designed to operate in the wavelength range from 1200 to 1600 nm, depending on the aluminium concentration of the collector region adjacent to the barrier.

4.4 RTD-EAM design and operation principle

A waveguide electroabsorption modulator requires light confinement and propagation along the region with the largest electrical field enhancement. In the case of an RTD-EAM waveguide, the DBQW structure is embedded in the core of the semiconductor optical waveguide. The waveguide core is sandwiched between two layers of a larger bandgap (lower refractive index) material, forming the cladding layers, see figure 4.5(a). Disregarding the larger bandgap layers, this new structure is identical to an RTD with two large spacer layers on each side of the DBQW. Finally, highly doped layers of a lower bandgap material are added for ohmic contacts.

Figure 4.5(b) schematically shows the unipolar channel ridged semiconductor optical waveguide structure configuration and the corresponding epilayer Γ-conduction band-edge and refractive-index profiles. This device is called a resonant tunnelling diode optical waveguide (RTD-OW). This ridged channel waveguide channel confines the butt-coupled light along the directions that contain the tunnelling layers due to the refractive-index distribution; therefore, allowing a more extended light interaction length (up to 100-fold larger than the light wavelength) within the region of the largest electric field magnitude.

The RTD-OW operation as an EAM can easily be understood if we consider the expected current–voltage characteristic of a similar unipolar waveguide structure where the DBQW structure is not included, i.e., a waveguide device with almost linear *I–V* characteristic that does not have an NDC region [8]. The DBQW included in the waveguide core strongly modifies the waveguide *I–V*, leading to a strongly nonlinear N-shape current–voltage characteristic. As a result, a nonlinear electric field distribution across the waveguide core is achieved. The electric field is strongly dependent on the bias voltage due to the electron accumulation close to the emitter barrier and due to the creation of a depletion region on the spacer layer of the collector side. Figure 4.6 schematically shows the light absorption in an

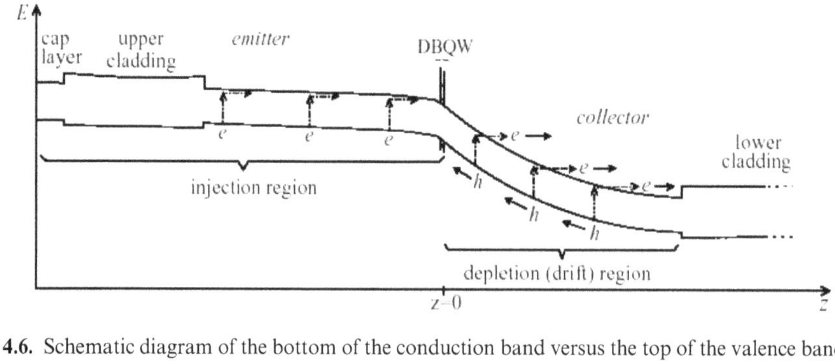

Figure 4.6. Schematic diagram of the bottom of the conduction band versus the top of the valence band versus distance illustrating the change in optical absorption induced by Franz–Keldysh effect in an RTD-OW biased around the valley point.

RTD-OW biased around the valley point. When the device is biased in the valley region, the absorption of light with energy slightly below the bandgap of the waveguide core is strongly enhanced. This results from the Franz–Keldysh effect in the depleted region of the collector side [6, 7].

Since a slight increase in the bias voltage can switch the RTD-OW operation point from the peak to the valley region, a large increase in the electric field across the collector depletion region can be achieved, resulting, through the Franz–Keldysh effect, in a pronounced absorption band-edge broadening effect. This leads to a substantial increase in the propagation losses in the waveguide at photon energies close to but smaller than the bandgap energy of the waveguide's core material. This unique property can be exploited to implement either electroabsorption modulators (EAM), semiconductor-based optical (intensity) modulators, or waveguide photodetectors (as discussed in chapter 5). Moreover, with appropriate design, it is possible to implement interferometric intensity modulators, as illustrated in figure 4.1, based on the electro-refraction effect.

We need to estimate the electric field enhancement induced by the peak-to-valley switching to evaluate the absorption change. For a given current density J, the magnitude of the electric field across the device depletion region as a function of the position, z, in the direction of the current (follow perpendicular to e-epilayers plane), $F(z)$, which can be obtained from Poisson's equation, as:

$$F(z) = F_{0,\,J} - \frac{e}{\varepsilon}(N_d^+ - J)z \tag{4.1}$$

where $F_{0,\,J}$ represents the electric field at the interface between the collector barrier and the collector spacer layer (where $z = 0$). The voltage drop across the collector region, ΔV_{PV}, due to the peak-to-valley switching is then given by:

$$\Delta V_{\mathrm{PV}} = F_{0,\,J}W_{\mathrm{dep}} + \frac{W_{\mathrm{dep}}^2}{2\varepsilon v_{\mathrm{sat}}}\Delta J_{\mathrm{PV}} \tag{4.2}$$

where $\Delta F_{0,\,J}$ represents the electric field enhancement due to the peak-to-valley switching at the interface between the collector barrier and the collector spacer layer, v_{sat} is the carrier saturation velocity and it is assumed that the thickness of the depletion (or drift) region, W_{dep}, does not change. From the relation (Equation 4.2), we can estimate the electric field enhancement, ΔF_{PV}, as a function of the peak-to-valley voltage difference as [8, 9]:

$$\Delta F_{\mathrm{PV}} \cong \frac{\Delta V_{\mathrm{PV}}}{W_{\mathrm{dep}}} + \frac{W_{\mathrm{dep}}}{2\varepsilon v_{\mathrm{sat}}}\Delta J_{\mathrm{PV}} \tag{4.3}$$

The Franz–Keldysh absorption coefficient change because of the electric field enhancement due to the peak-to-valley switching is given by [6]:

$$\Delta\alpha\left(\frac{hc}{\lambda},\, \Delta F_{\mathrm{PV}}\right)_{\mathrm{PV}} = \alpha\left(\frac{hc}{\lambda},\, F_{\mathrm{V}}\right)_{\mathrm{V}} - \alpha\left(\frac{hc}{\lambda},\, F_{\mathrm{P}}\right)_{\mathrm{P}} \tag{4.4}$$

where $\alpha\left(\frac{hc}{\lambda}, F\right)$ is given, in the weak field approximation, by the electroabsorption coefficient resultant from the Franz–Keldysh effect, see section 3.3.2.

At a given photon energy, $E = \frac{hc}{\lambda}$, the absorption change induced by the electric field enhancement due to the peak-to-valley switching can be estimated as:

$$\Delta\alpha\left(\frac{hc}{\lambda}, \Delta F_{\mathrm{PV}}\right)_{\mathrm{PV}} \cong \alpha\left(\frac{hc}{\lambda}, \Delta F_{\mathrm{PV}}\right)_{\mathrm{V}} - \alpha_0\left(\frac{hc}{\lambda}\right) \tag{4.5}$$

where $\alpha_0\left(\frac{hc}{\lambda}\right)$ corresponds to the absorption loss of the material at wavelength λ. For simplicity of analysis, we assume that $\alpha\left(\frac{hc}{\lambda}, F_{\mathrm{V}}\right)_{\mathrm{V}} \gg \alpha\left(\frac{hc}{\lambda}, F_{\mathrm{P}}\right)_{\mathrm{P}} \cong \alpha_0\left(\frac{hc}{\lambda}\right)$ because $F_{\mathrm{V}} \cong \Delta F_{\mathrm{PV}} \gg F_{\mathrm{P}}$.

The Franz–Keldysh absorption band-edge shift resultant from the peak-to-valley switching can be estimated using the relation:

$$\Delta\lambda_g \cong \frac{\lambda_g^2}{hc}\left(\frac{eh}{8\pi^2 m_r}\right)^{\frac{2}{3}} \Delta F_{\mathrm{PV}}^{\frac{2}{3}} = \frac{\lambda_g^2}{hc}\left(\frac{eh}{8\pi^2 m_r}\right)^{\frac{2}{3}}\left(\frac{\Delta V_{\mathrm{PV}}}{W_{\mathrm{dep}}} + \frac{W_{\mathrm{dep}}}{2\varepsilon v_{\mathrm{sat}}}\Delta J_{\mathrm{PV}}\right)^{\frac{2}{3}} \tag{4.6}$$

The measurement of $\Delta\lambda_g$ enables evaluation of the electric field change, ΔF_{PV}.

It is possible to design RTD-OWs with a sharp current drop across the NDC and achieve a peak-to-valley switching with applied high frequency voltage AC signals ($\gg$10 GHz) with an amplitude less than 1 V. In this case, the voltage drop across the depletion region is given by [10, 11]:

$$\Delta V_{\mathrm{PV}} = V_{\mathrm{P}}\left(1 - \frac{J_{\mathrm{V}}}{J_{\mathrm{P}}}\right) \tag{4.7}$$

Assuming that at the peak bias voltage a significant portion of the core is depleted, high-speed electric field switching can be achieved, leading to high-frequency modulation of the waveguide optical transmission loss. In this situation, the RTD-OW operates as a RTD-EAM, i.e., the modulation depth of the device depends mainly on the overlap between the electric field in the collector depleted region and the optical mode. In essence, the increase in the magnitude of the electric field is determined by the characteristics of the NDC region.

The band-edge shift described previously results in an increase in the light-wave propagation loss, i.e., the peak-to-valley switching leads to a change in the intensity of the waveguide output. The modulator transmission coefficient, dependent on the applied voltage, is:

$$T(V) \propto \exp\left[-\gamma_f \cdot \alpha(V) \cdot L\right] \tag{4.8}$$

where $\alpha(V)$ is the waveguide core material absorption coefficient under a given bias voltage, V, L is the length of the waveguide section under electric contact (see figures 4.2 and 4.5), and γ_f is the optical filling factor corresponding to the fraction of the optical power guided in the depleted region of the waveguide.

For simplicity of analysis, neglecting the reflections and couplings losses at the facets of the waveguide, the key figure-of-merit of the modulator is the modulation depth, or the modulator on/off ratio (or contrast ratio), $R_{\mathrm{on/off}}$, due to the peak-to-valley switching, $R_{\mathrm{on/off}} = R_{\mathrm{VP}}$, which is given by:

$$R_{\mathrm{on/off}} = \frac{P_{\mathrm{out,on}}(V_{\mathrm{on}} = V_{\mathrm{P}})}{P_{\mathrm{out,off}}(V_{\mathrm{off}} = V_{\mathrm{V}})} = \frac{T(V_{\mathrm{P}})}{T(V_{\mathrm{V}})} \tag{4.9}$$

In decibels,

$$R_{\mathrm{on/off}}(\mathrm{dB}) = 10\log\frac{T(V_{\mathrm{P}})}{T(V_{\mathrm{V}})} = 4.343\gamma_f[\alpha(V_{\mathrm{V}}) - \alpha(V_{\mathrm{P}})]\cdot L \tag{4.10}$$

that is:

$$R_{\mathrm{VP}} = R_{\mathrm{on/off}} \quad (\mathrm{dB}) \cong 4.343\gamma_f\Delta\alpha\left(\frac{hc}{\lambda}, \Delta F_{\mathrm{PV}}\right) \times L \tag{4.11}$$

Since the modulator insertion loss of the transmission (on) state increases with L as:

$$\Upsilon_{\mathrm{in}} = \frac{P_{\mathrm{in}} - P_{\mathrm{on}}(V_{\mathrm{P}})}{P_{\mathrm{in}}} = 1 - T(V_{\mathrm{P}}) = 1 - \exp\left[-\gamma_f\alpha(V_{\mathrm{P}}) \cdot L\right] \tag{4.12}$$

an optimised design is a trade-off between the extinction ratio and the insertion loss.

Another fundamental figure-of-merit in the design of EAMs is the absorption coefficient change per unit of applied voltage:

$$\frac{\Delta\alpha}{\Delta V} = \frac{\alpha(V_{\mathrm{off}}) - \alpha(V_{\mathrm{on}})}{V_{\mathrm{off}} - V_{\mathrm{on}}} \tag{4.13}$$

These two figures of merit can be combined as the modulator on/off ratio (in dB) per unit of applied voltage, given by:

$$\frac{R_{\mathrm{on/off}}(\mathrm{dB})}{\Delta V} = 4.343\frac{\alpha(V_{\mathrm{V}}) - \alpha(V_{\mathrm{P}})}{\Delta V_{\mathrm{VP}}} \tag{4.14}$$

In our previous work, the RTD-EAM concept was demonstrated at 900 nm using the AlGaAs/GaAs material system, and at 1550 nm with InGaAlAs/InP. The quaternary compound InGaAlAs and InGaAsP lattice matched to InP can be used to implement RTD-EAMs operating at wavelengths within the 1300 and 1550 nm optical windows, wavelengths where the standard single-mode optical fibres have zero dispersion and lowest loss, respectively. In what follows, we describe the work on RTD-OW EAMs at optical communication windows around 900 nm and 1550 nm.

4.5 GaAs RTD-EAM operation at 900 nm

4.5.1 Device implementation and low frequency characterisation

For operation in the spectral region around 900 nm, the RTD-OW electroabsorption modulator (RTD-EAM) structure consists of an AlGaAs/GaAs waveguide structure grown on semi-insulating GaAs wafers: the waveguide core is made of GaAs (Si doped 2×10^{16} cm^{-3}), and two $Al_{0.33}Ga_{0.67}As$ (Si doped 2×10^{18} cm^{-3}) layers are employed as cladding layers of the waveguide (the $Al_{0.33}Ga_{0.67}As$ is a direct bandgap compound alloy). The AlAs/GaAs/AlAs DBQW was embedded in the middle of 1 µm thick GaAs core layer to allow easy coupling of the light input and output (butt-coupling). The DBQW comprises a 7 nm thick GaAs quantum well sandwiched between 1.4 nm thick AlAs barriers. A detailed description and fabrication of AlGaAs/GaAs structures can be found in [8]. A diagram of a typical GaAs/AlGaAs ridge waveguide is shown in figure 4.7; the conduction band-edge E_C and refractive index *RI* profiles in the growth direction are also represented.

The refractive index difference between the core and cladding layers is 0.224 at 900 nm, which is sufficient to obtain highly efficient light confinement using relatively thin cladding layers. The thickness of the top cladding layer is 300 nm, which is twice the reciprocal of the first mode exponential decaying factor, to achieve a low series resistance. Because the waveguide core and the substrate have similar real refractive indices, the bottom cladding layer is made twice as large as the top cladding—i.e., 600 nm thick, and therefore separates the core from the substrate and acts as an optical isolation layer—to significantly reduce radiation leakage into the substrate. Figure 4.8 shows (a) the top view of an RTD-OW die and (b) a packaged device.

A typical current–voltage (*I*–*V*) characteristic of the AlGaAs/GaAs devices is shown in figure 4.9.

The optoelectronic characterisation employed light from a tunable Ti:sapphire laser and includes measurements of: (i) the waveguide propagation loss, determined by sequential cleaving method; (ii) the waveguide absorption band-edge; and (iii) the transmission spectrum. For the case of waveguides with widths above 3 µm, the propagation loss at 908 nm was below 5.5 cm^{-1}. The core band-edge was around 890 nm. The RTD-EAM waveguide transmission spectra for devices with active areas around 800 µm^2 (4 µm × 200 µm) are shown in figure 4.10 at zero bias, when biased slightly below the peak, and at the valley point. The devices were direct current (DC) biased outside the NDC region to avoid entering the self-oscillation region.

As seen in figure 4.10, by changing the quiescent point from the peak to the valley region, a sharp drop in the waveguide transmission is observed in the wavelength

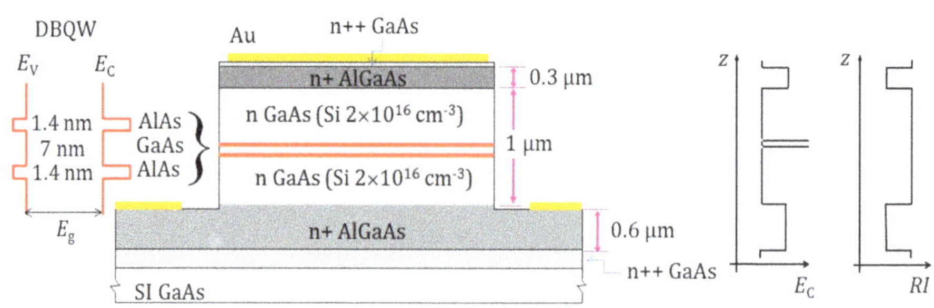

Figure 4.7. A ridge GaAs/AlGaAs DBQW waveguide epilayer structure; the conduction band-edge and refractive-index profiles in the growth direction are also shown.

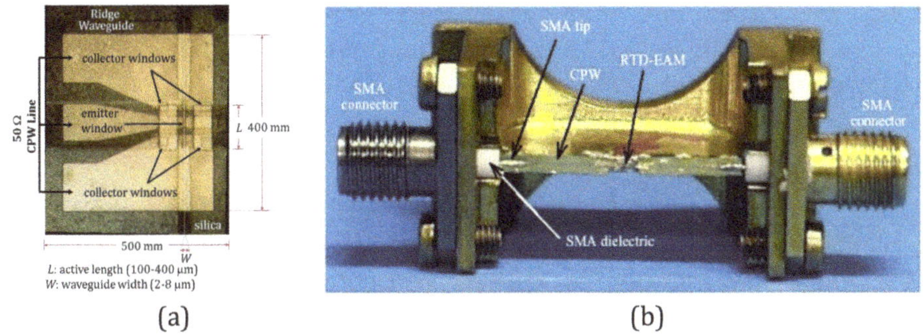

Figure 4.8. Photograph of: (a) a RTD-EAM die, here L represents the device electrical active length (contact length), which with the waveguide width defines the device's active area; and (b) a connectorised prototype package device.

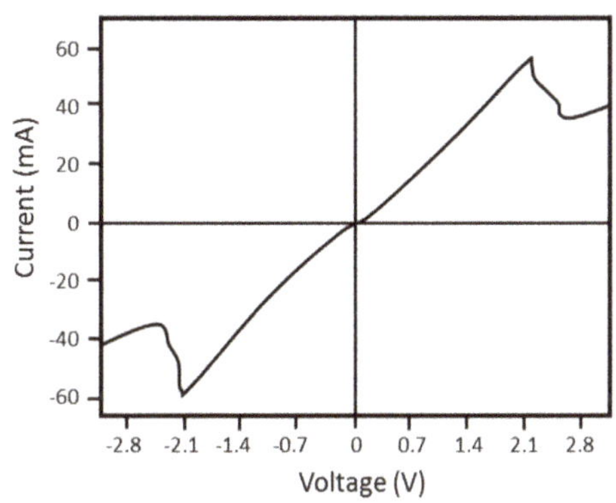

Figure 4.9. Typical full I–V characteristic AlGaAs/GaAs with a 7 nm GaAs and 1.7 nm AlAs barriers. The peak current was around 13.5 kA cm^{-2} and the peak-to-valley current ratio ranged from 1.4 to 1.8. The peak voltage values were 1.5–3 V, and the valley voltage was 1.8–3.5 V, depending mostly on the area of the devices.

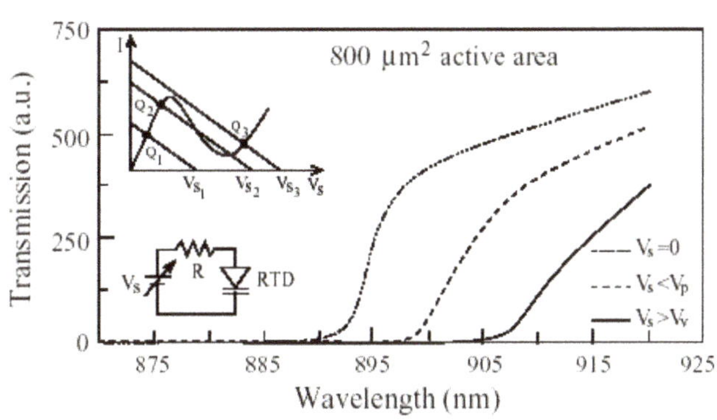

Figure 4.10. Optical transmission spectra at different dc bias points: no bias, biased slightly below the peak and at the beginning of the valley region for 4 μm × 200 μm devices, respectively.

range of 890–920 nm. The respective measured Franz–Keldysh absorption band-edge wavelength shift was ~12 nm, which compares to the 9 nm value estimated from equation (4.6), which takes into consideration the approximations used in [9].

4.5.2 Optical modulation under applied external signals

In this subsection, we describe the RTD-OW operation as an electroabsorption modulator. The device is DC biased in the second positive differential conductance region (where the detrimental thermal effects are less pronounced). The modulation depth was evaluated by driving the devices with an RF signal applied through a wide bandwidth bias-T (square voltage signal with a peak-to-peak amplitude slightly higher than the peak-to-valley voltage difference, ΔV_{VP}), see the inset of figure 4.11. The optical modulation depth response as a function of the wavelength due to the transition between the two positive PDC regions is shown in figure 4.11: modulation depths up to 13 dB at 908 nm were achieved, and modulation depths up to 18 dB were observed in the case of 4 μm × 400 μm ridges waveguides.

Figure 4.12(a) shows examples of direct modulation at 950 MHz and 16 GHz. In both cases, the amplitude of the driving signals was slightly larger than $\Delta V_{VP} \cong 0.4$ V. Optical modulation depths as large as 11 dB were achieved at 16 GHz [9, 12]. The 16 GHz response in the spectral domain is shown in figure 4.12(b). This provides an estimated modulation depth for the devices analysed here, showing their potential for high-bandwidth applications.

The modulation efficiency characterised by the bandwidth-to-drive-voltage ratio, defined as the ratio of the operation bandwidth to the operating voltage ($B/\Delta V$) for a 10 dB modulation depth, was above 40 GHz V^{-1} [9].

4.5.3 Optical modulation enabled by self-sustained oscillations

Whenever an RTD (RTD-EAM) is part of a resonant tank circuit, auto-oscillations can follow once the DC bias point is set in the NDC region. In this subsection, we

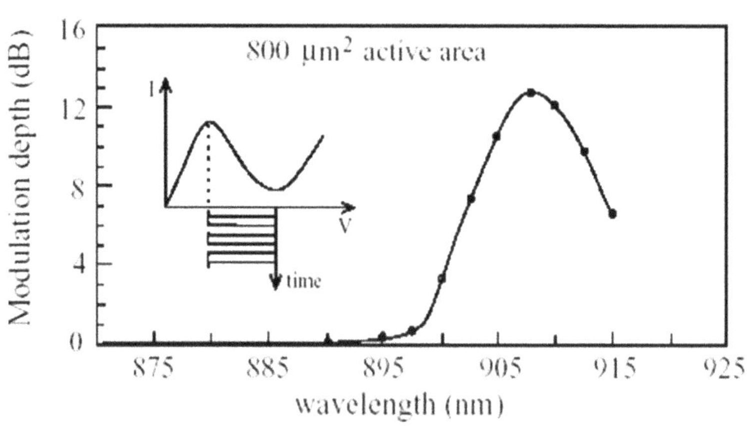

Figure 4.11. Modulation depth as a function of the operating wavelength due to peak-to-valley switching induced by a square voltage waveform. The devices were dc biased in the valley region.

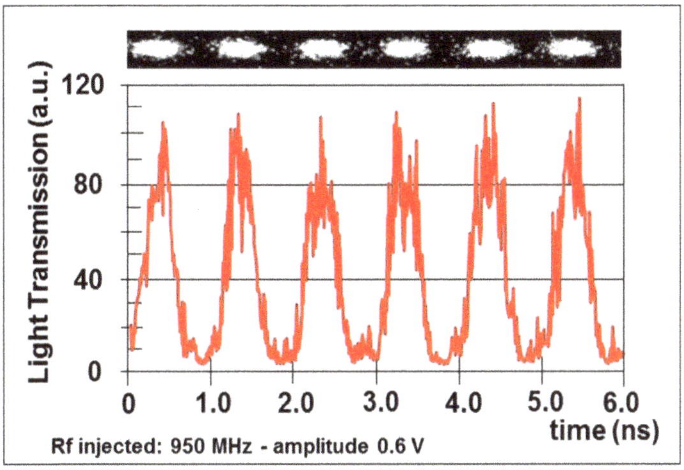

(a)

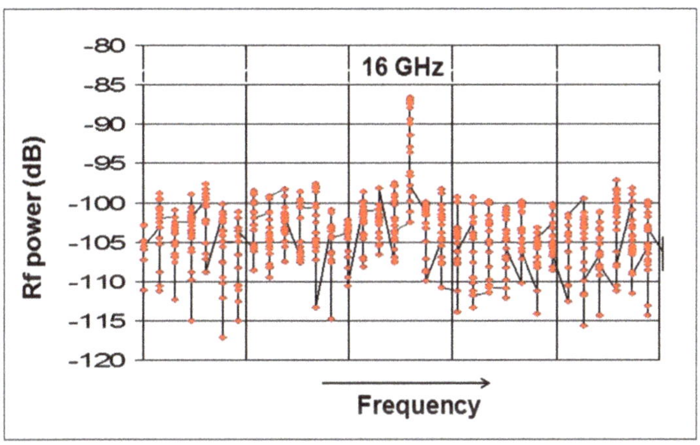

(b)

Figure 4.12. Direct modulation: (a) at 950 MHz and (b) at 16 GHz. The measured modulation depth was 11 dB (λ = 908 nm) at 16 GHz and the bandwidth-to-drive-voltage ratio was greater than 40 GHz V^{-1}.

analyse the case of RTD-EAM devices connected to a transmission line via a bias-T. This configuration forms an RTD-EAM transmission line relaxation oscillator [13]. Figure 4.13 shows a schematic of the RTD-EAM relaxation oscillator circuit where the length of the coaxial transmission line determines the frequency of oscillation [10]. The RTD-EAM and the CPW transmission line plus the connector represent the connectorised prototype package device shown in figure 4.8(b). The inductance associated with the length of the transmission line and the device capacitance define a resonant tank circuit. When the RTD-EAM is DC biased in the NDC region, self-sustained oscillations can be achieved. The relaxation oscillation frequency can be tuned by changing the dc bias voltage. This changes the RTD-EAM intrinsic impedance that depends on the applied voltage, forming a voltage-controlled oscillator.

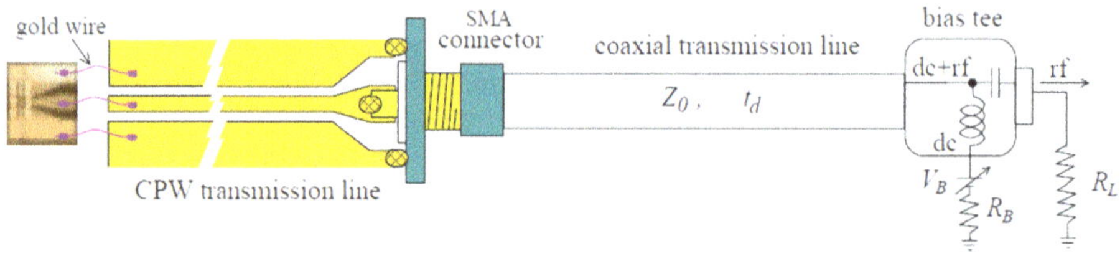

Figure 4.13. Schematic representation of an RTD-EAM transmission line relaxation oscillator.

Additionally, the frequency of oscillation can be tuned by the optical power coupled to the RTD-EAM. This behaviour enables the implementation of RTD-based optically controlled oscillators. This frequency tuning effect is mainly related to the creation of charge carriers in the depletion region, which reduces the series resistance of the RTD-EAM, changing the operating point in the NDC region. This in turn changes the device impedance.

As discussed earlier, when a RTD-EAM connected to a bias-T through a transmission line (coaxial cable) is DC biased in the NDC region, the self-oscillations modulate the optical output, as shown in figure 4.14. The oscillation frequencies are determined essentially by the electrical length of the transmission line and the capacitance of the devices.

Additional studies of the response of connectorised prototype package devices (see figure 4.8(b), where the transmission line corresponds to the CPW transmission line used to connect the device to the SMA connector) have shown that AlGaAs/GaAs RTD-EAMs under external excitation can produce pulses of light as short as 30 ps and optical modulation up to 14 GHz induced by the device self-oscillation [13]. Analysis of these results suggests that the frequency was limited by the electronic packaging of the devices employed in the characterisation.

The observed response of the AlGaAs RTD-EAM indicate that these devices have potential for application in local area networks, primarily to electrically control guided-wave optical signals in the wavelength range of 880 nm to 1100 nm. This includes the development of new types of waveguide intensity modulators, directional couplers and optical switches.

The high-speed self-sustained relaxation oscillation operating mode of RTD-EAMs can be applied in clock extraction circuits, for optical pulse generation and for de-multiplexing in optical time division multiplexed systems. Furthermore, because light changes the impedance of the resonant tank circuit, the RTD-EAM can also act as a radio frequency oscillator, with its frequency being optically controlled—an optoelectronic-voltage-controlled oscillator.

4.6 InGaAlAs/InP RTD-EAM operation at 1550 nm

4.6.1 Device structure and low frequency characterisation

The AlGaAs/GaAs RTD-EAM device and its performance guided our work towards demonstrating a conceptually similar device operating at the fibre third optical transmission window at 1550 nm, where single-mode optical fibres have the

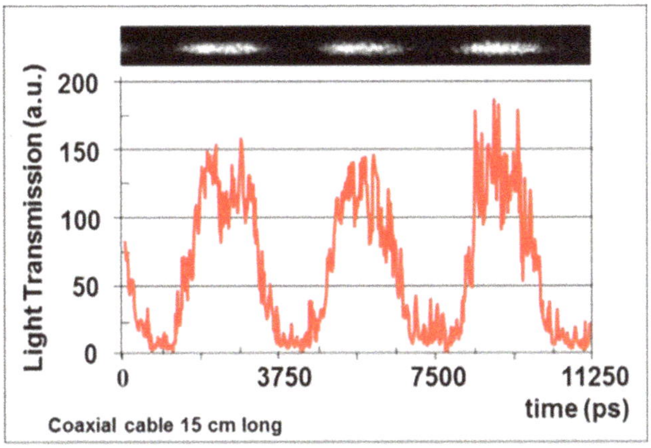

(a)

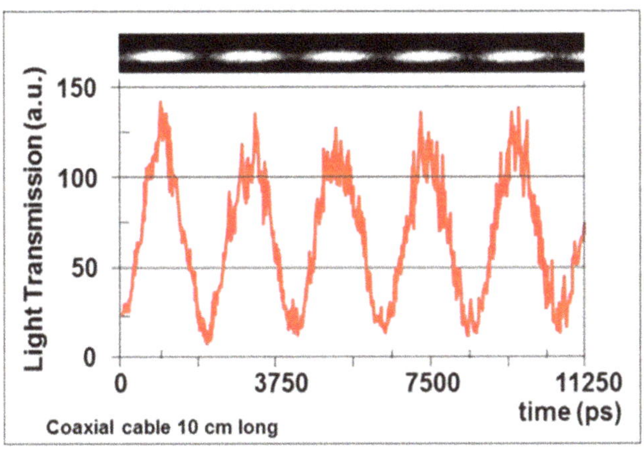

(b)

Figure 4.14. Optical responses of RTD-EAM transmission line relaxation oscillators with coaxial transmission line lengths of (a) 15 cm and (b) 10 cm.

lowest loss. Several semiconductor alloys can be used for EAMs operating at 1550 nm, e.g., InGaAlAs and InGaAsP. These materials have favourable properties for high-performance RTD-based devices, such as a larger electron mobility and lower electron effective mass than GaAs-based compound semiconductors. These properties lead to RTD-EAMs with higher performance, namely increased speed and modulation depth. The InGaAlAs/InP RTD-EAMs show larger peak current densities and PVCR, and smaller peak voltage as compared to AlGaAs/GaAs RTD-EAMs. Furthermore, the quaternary system InGaAlAs is lattice matched to InP, enabling operation also at 1300 nm (another telecommunications optical window) [8].

The epilayer structure of the InGaAlAs/InP RTD-EAM for operation at 1550 nm is shown schematically in figure 4.15, which includes a waveguide structure with 1 um

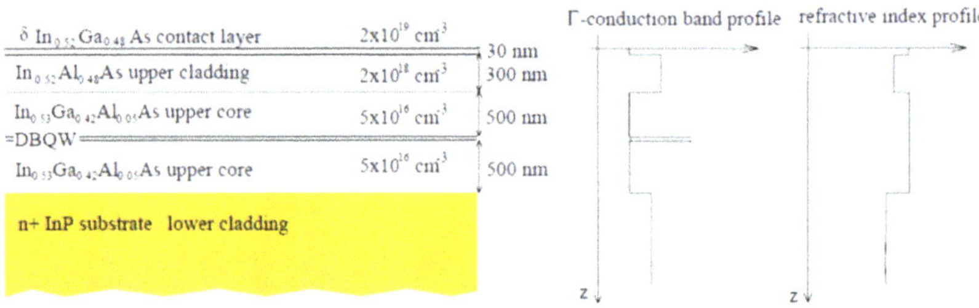

Figure 4.15. InGaAlAs/InP RTD-EAM structure, and the Γ-valley and refractive-index profiles.

$In_{0.53}Ga_{0.42}Al_{0.05}As$ thick core with a 300 nm $In_{0.52}Al_{0.48}As$ top cladding layer where the InP substrate works as the bottom cladding layer. Figure 4.15 also shows the Γ-valley conduction band profile and the refractive-index profile of the waveguide structure. The waveguide core consists of two $In_{0.53}Ga_{0.42}Al_{0.05}As$ layers (refractive index of 3.56), each with a thickness of 500 nm. The DBQW sits between these two layers. The $In_{0.53}Ga_{0.42}Al_{0.05}As$ bandgap energy is 0.826 eV, corresponding to an absorption band-edge wavelength close to 1500 nm to allow operation at around 1550 nm when under bias in the RTD-EAM peak voltage. The top cladding consists of a thick 300 nm layer of $In_{0.52}Al_{0.48}As$ (refractive index, $n = 3.24$). In this configuration, the InP substrate acts as the bottom cladding region because the refractive index of InP at 1550 nm ($n =$ 3.17) is considerably lower than the refractive index of the $In_{0.53}Ga_{0.42}Al_{0.05}As$ core layer. As discussed previously, the 300 nm thickness of the top cladding layer corresponds to approximately twice the decay length of the fundamental optical mode in this region. A delta doped InGaAs layer was included as the top caping layer for lower resistance ohmic contacts. The lower contact is formed on the highly doped InP substrate. A more detailed description of the wafer structure is given in [8].

The typical current–voltage characteristic of $4\ \mu m \times 100\ \mu m$ InGaAlAs/InP RTD-EAM is presented in figure 4.16, showing a peak-to-valley current ratio (PVCR) of around 4. Most of the RTD-EAMs characterised were ridges waveguides that were 4 μm wide with 200 μm active lengths (defined as the top contact length), the waveguides were 500 μm long in total. These devices showed valley-to-peak voltage differences ΔV_{VP} up to 0.8 V, with peak-to-valley current density differences ΔJ_{PV} as high as 10 kA cm^{-2}. For comparison, typical GaAs/AlAs devices show $\Delta V_{VP} \approx 0.4$V and $\Delta J_{PV} \approx 5$ kA cm^{-2}.

Impedance measurements on-wafer in the frequency range 45 MHz to 18 GHz for all values of bias voltage indicates the InGaAlAs RTD-EAM small signal equivalent circuit can be modelled as a capacitance C in parallel with a nonlinear resistor R_d, in series with a resistance R_S, see figure 4.17. The series inductance was found to be negligible [14].

The device switching time can be estimated using [10]:

$$t_R = 4.4C_V\left(\frac{\Delta V_{VP}}{\Delta J_{PV}}\right) \tag{4.15}$$

where C_V is the device capacitance per unit area (assumed to be given by $C_V \approx \varepsilon / W_{dep}$). The average capacitance C_V and the shunt resistance R_d of these

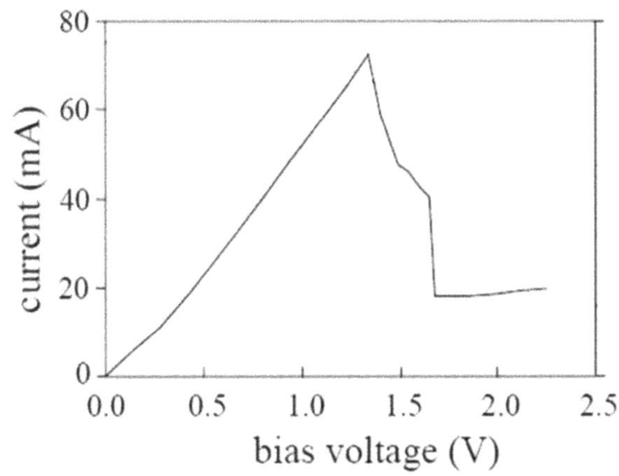

Figure 4.16. The typical current–voltage characteristics of 4 μm × 100 μm InGaAlAs/InP RTD-EAM.

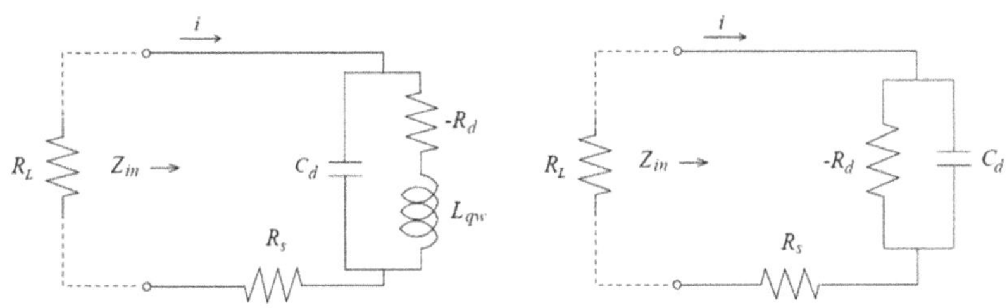

Figure 4.17. Small signal equivalent circuits of a RTD-EAM (a) considering the quantum inductance and (b) ignoring the quantum inductance L_{qw} (i.e. ignoring the time necessary to build up the charge in the quantum well to change the current), which is acceptable for the operating frequencies under consideration.

devices, when biased in the NDC region, were estimated to be 1.5 pF and −15 Ω, respectively, assuming $R_S = 3\ \Omega$. This gives a switching time of $t_R = 6.6$ ps [14].

As in the case of the GaAs-based devices, the DC/low frequency electroabsorption response of the waveguide was characterised as a function of the applied DC voltage with no applied high-frequency signal. The DC bias voltage range avoided biasing in the NDC region to avoid self-oscillations. Light from a Tunics diode laser, tuneable in the wavelength range 1480–1580 nm, was fibre butt-coupled to the waveguide. The waveguide light output was collected in a fibre butt-coupled arrangement. An optical power meter and a high-bandwidth photodetector were used to detect the light output. For high-speed characterisation, a fibre pigtailed 45 GHz bandwidth Schottky InGaAs photodetector (NewFocus) was used.

The transmission spectra of the typical InGaAlAs/InP RTD-EAM for bias point around zero volts, and when biased close to the peak and the valley voltages, are shown in figure 4.18. The measured band-edge shift was 43 nm, which compares well with the estimated value of 46 nm, equation (4.6). This indicates that an electrical field of about 38 kV cm^{-1} is needed to achieve the observed band-edge broadening effect.

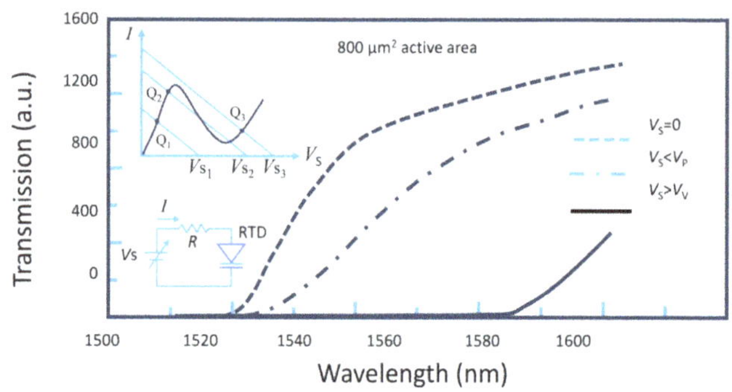

Figure 4.18. InGaAlAs RTD-EAM transmission spectra in the wavelength range 1500–1580 nm, with the applied voltage as a parameter.

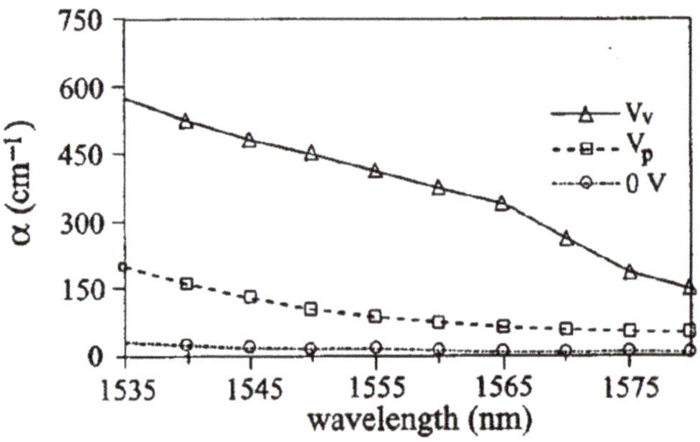

Figure 4.19. Absorption coefficient as a function of the wavelength with the applied voltage as a parameter: at zero bias, and at peak and valley points.

The absorption coefficient spectrum was calculated from the experimental data for the same biasing conditions as figure 4.18: at zero bias, when biased around the peak and valley regions. The results are shown in figure 4.19.

The variation of the absorption coefficient induced by the peak-to-valley transition, $\Delta\alpha_{VP} = \alpha_V - \alpha_P$, was calculated and is presented in figure 4.20.

The level of electroabsorption can be estimated by calculating the change in the absorption coefficient. This is given by the absorption coefficient change due to peak-to-valley switching, $\Delta\alpha_{VP}$, over the absorption coefficient on the on-state α_P (in the case of the RTD-EAM, at peak bias voltage, V_P), $\Delta\alpha_{VP}/\alpha_P$, as a function of the wavelength, as shown in figure 4.21.

The modulation depth as a function of the light wavelength because of the transition between the two PDC regions when a square-wave signal is applied is shown in figure 4.22. The peak-to-peak voltage was $\Delta V_{VP} \approx 0.8$V. The devices were DC biased avoiding self-oscillations. The DC bias was set in the valley region because of the smaller thermal effects in this region (due to the smaller operating

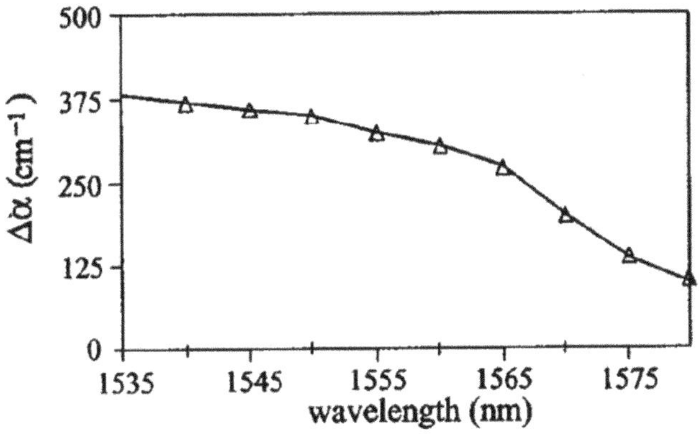

Figure 4.20. InGaAlAs RTD-EAM change in absorption coefficient spectrum in the wavelength range 1500–1580 nm when the RTD switches the voltage.

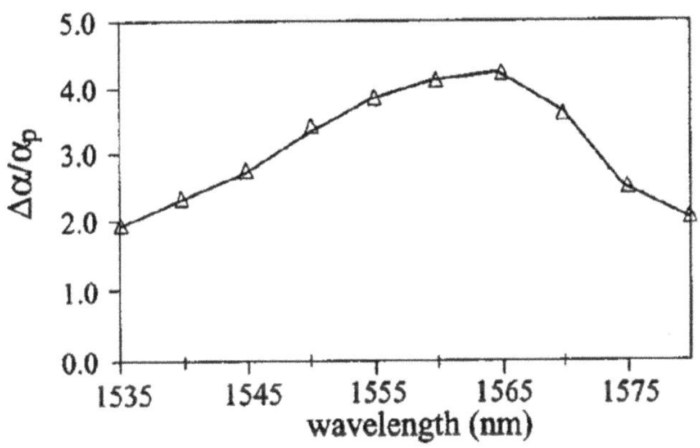

Figure 4.21. Ratio of change of absorption coefficient induced by electroabsorption, $\Delta\alpha_{VP}$, over the absorption coefficient on the on-state (at peak bias voltage) α_P.

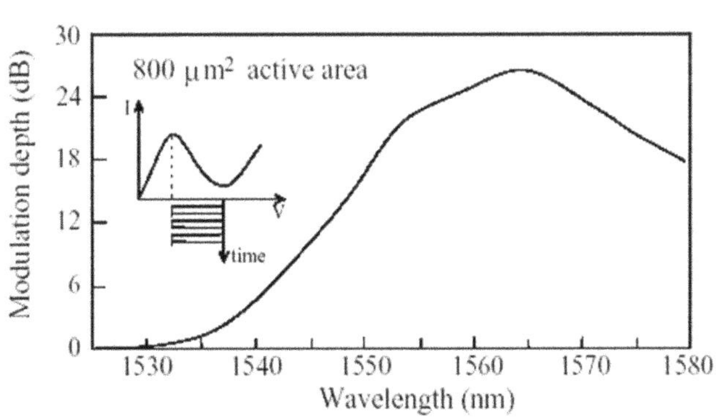

Figure 4.22. Modulation depth as a function of the wavelength.

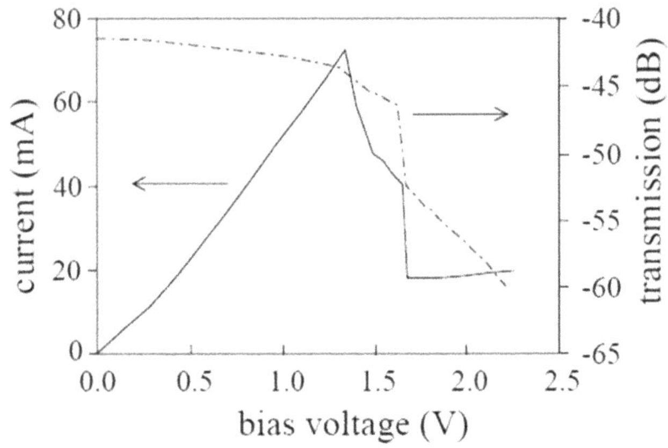

Figure 4.23. InGaAlAs/InP RTD-EAM structure, Γ-valley and refractive-index profiles. The waveguide transmission as a function of the bias voltage is also shown, a very sharp drop in the transmission in response to a minimal voltage change can be seen.

current than in the peak region). In this configuration, a low frequency modulation depth of 28 dB was measured for the case of a device with an active area of 800 μm^2 (corresponding to 4 μm ridge waveguides). These values are 10 dB larger than the modulation depths measured in AlGaAs/GaAs devices. The transmissive state propagation loss was estimated from the measurements to be around 5 dB.

The waveguide transmission coefficient was also measured through the full *I*–*V* for some devices. Figure 4.23 presents the transmission coefficient as a function of the applied bias voltage. This result shows a reduction of 5 dB in the transmission in the case of a 1 mV increase of the DC voltage at a bias point ~1.6 V. Effectively, this corresponds to an exceptionally large transmission change per unit of voltage.

4.6.2 RTD-EAM high-frequency characterisation

The high-frequency characterisation started by measuring the modulator response up to 3 GHz. Figure 4.24 plots the modulator response as a function of the DC bias voltage with the amplitude of the RF signal as a tunable parameter for a 3 GHz sinewave RF voltage signal. The *I*–*V* characteristic of the RTD-EAM is also shown in figure 4.24. The modulator high-frequency response increased ~15 dB when the device DC bias changed from the peak to the valley region, even for RF driving signals with an amplitude as low as 50 mV. This is a clear indication that the RTD-EAM can achieve impressive modulation depths, even in the cases of extremely low (≪1 V) amplitude voltage signals.

To test the RTD-EAM at frequencies higher than 10 GHz, a microwave synthesised signal generator with a maximum output of +20 dBm and a maximum frequency of 26 GHz was employed. The modulator response in the case of a 26 GHz driving signal with 7.7 dBm RF power is shown in figure 4.25. The photo-detected power increases more than 10 dB when the driving rf power increases from −220 dBm to +7.7 dBm. This is a clear indication that the device can reach modulation extinction ratios larger than 10 dB for low-power driving signals,

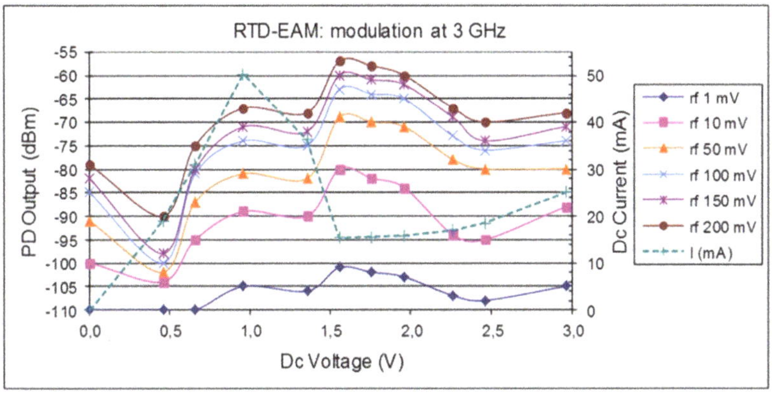

Figure 4.24. Modulator response as a function of the DC bias voltage when driven by 3 GHz rf signals, with injected amplitude as a parameter.

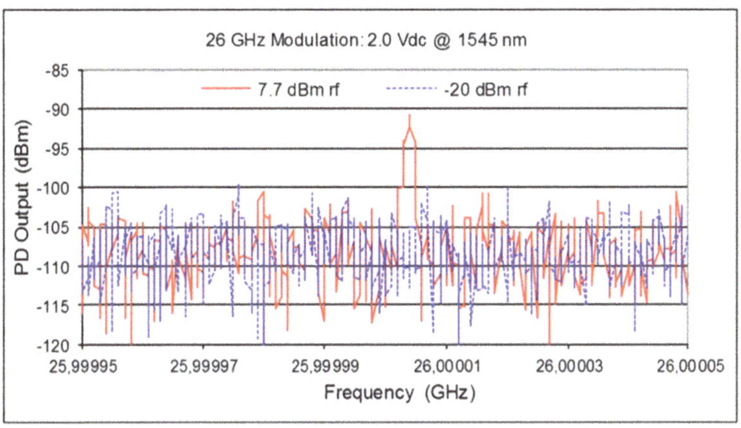

Figure 4.25. Spectra of the 26 GHz photo-detected signals for a modulator driving power of −20 and +7.7 dBm, respectively.

typically lower than 10 mW, as a consequence of the built-in electrical amplification effect provided by the RTD-EAM devices.

The amplification effect (due to the intrinsic NDC of RTD-EAMs) can substantially reduce the rf power required to achieve large modulation depths, e.g., 10 dB. This on-chip amplification can eliminate the need for the external rf amplifiers that are usually required to drive EAMs [1], and therefore provides practical solutions for low-power and compact high-speed modulators.

4.7 Conclusions

In optical communication applications, waveguide EAMs have been considered to encode digital information onto an optical signal. Waveguide EAMs have been used in optical communication applications to encode digital data onto an optical signal.

In this chapter, we covered the RTD-EAM and how the nonlinear current–voltage characteristic of the DBQW-RTD can efficiently switch rapidly the built-in electric field in an optical waveguide core. Incorporating a DBQW in the core of a semiconductor optical waveguide makes the waveguide current–voltage characteristic highly nonlinear, leading to a highly nonlinear built-in electric field distribution across the waveguide cross section. This allows EAMs operating with much less rf power to be implemented.

The electric field changes the optical absorption of the waveguide via the Franz–Keldysh effect, so that light with photon energy just below the waveguide core bandgap energy becomes strongly absorbed.

The RTD-EAM concept was first realised using an AlGaAs/GaAs material system for operation at wavelengths around 900 nm. The RTD-EAM concept was then transferred to the InGaAlAs/InP material system, which had its bandgap tuned to the optical communications 'C' band, wavelengths around 1550 nm. The InP-based RTD-EAMs were characterised up to 26 GHz, showing up 10 dB modulation depth at a low-power driving of around 10 mW. The experiments indicate that when DC is biased to the NDC region, incorporating the RTD with the EAM waveguide structure can obliviate the need for an external rf amplifier, considerably reducing the system's complexity.

The intrinsic high speed of the tunnelling process in RTDs, which has already allowed the demonstration of fundamental oscillations over 1.9 THz [15], and the possibility of combination with other optoelectronic and electronic devices give the RTD-EAM high potential for use in the next generation of information technology systems, in particular for more energy efficient high-speed communication and signal processing implementations.

References

[1] Wakita K 1998 *Semiconductor Optical Modulators* (Boston, MA: Springer)

[2] Zhu N H *et al* 2018 Directly modulated semiconductor lasers *IEEE J. Sel. Top. Quantum Electron.* **24** 1500219

[3] Eldada L 2004 Optical communication components *Rev. Sci. Instrum.* **75** 575–93

[4] Wooten E L *et al* 2000 A review of lithium niobate modulators for fiber-optic communications systems *IEEE J. Sel. Top. Quantum Electron.* **6** 69–82

[5] Bazzan M and Sada C 2015 Optical waveguides in lithium niobate: recent developments and applications *Appl. Phys. Rev.* **2** 040603

[6] Chuang S L 2010 *Physics of Optoelectronic Devices* (New York: Wiley)

[7] Tharmalingam K 1963 Optical absorption in the presence of a uniform field *Phys. Rev.* **130** 2204–6

[8] Figueiredo J M L 2000 Optoelectronic properties of resonant tunnelling diodes *PhD Thesis* Universidade of Porto http://userweb.eng.gla.ac.uk/charles.ironside/JLFThesis.pdf

[9] Figueiredo J M L, Ironside C N and Stanley C R 2001 Electric field switching in a resonant tunneling diode electroabsorption modulator *IEEE J. Quantum Electron.* **37** 1547–52

[10] Mcmeekin S G, Taylor M R S, Vogele B, Stanley C R and Ironside C N 1994 Franz–Keldysh effect in an optical wave-guide containing a resonant-tunneling diode *Appl. Phys. Lett.* **65** 1076–8

[11] McMeekin S G, Taylor M R S and Ironside C N 1996 Optical modulation with a resonant tunnelling diode *IEE Proc., Optoelectron.* **143** 12–6

[12] Figueiredo J M L, Boyd A R, Stanley C R, Ironside C N, McMeekin S G and Leite A M P 1999 Optical modulation at around 1550 nm in an InGaAlAs optical waveguide containing an InGaAs/AlAs resonant tunneling diode *Appl. Phys. Lett.* **75** 3443–5

[13] Figueiredo J M L, Stanley C R, Boyd A R, Ironside C N, McMeekin S G and Leite A M P 1999 Optical modulation in a resonant tunneling relaxation oscillator *Appl. Phys. Lett.* **74** 1197–9

[14] Alkeev N V, Lyubchenko V E, Ironside C N, Figueiredo J M L and Stanley C R 2000 Superhigh-frequency characteristics of optical modulators on the basis of InGaAlAs resonance-tunnel heterostructures *J. Commun. Technol. Electron.* **45** 911–4

[15] Maekawa T, Kanaya H, Suzuki S and Asada M 2016 Oscillation up to 1.92 THz in resonant tunneling diode by reduced conduction loss *Appl. Phys. Express* **9** 4

IOP Publishing

Resonant Tunnelling Diode Photonics Devices and Applications (Second Edition)

Charlie Ironside, Bruno Romeira and José Figueiredo

Chapter 5

Resonant tunnelling diode-photodetector (RTD-PD)

Semiconductor light sensors that incorporate double barrier quantum wells (DBQW) which take advantage of the N-shape current–voltage characteristics and the associated built-in wide-bandwidth electrical gain have nonlinear photocurrent–voltage and photocurrent–optical power (responsivity) relations. This allows them to achieve responsivities that are much superior to photoconductive detectors, such as PIN photodiodes (PIN-PDs) and avalanche photodiodes (APDs).

5.1 Introduction

In addition to the conventional high-frequency microwave and terahertz applications [1–3], double barrier quantum well resonant tunnelling diode (DBQW-RTD) based structures made of III–V semiconductor compounds have been considered for sensing physical quantities such as temperature, pressure and light [4–20]. Adding light sensitive layers to an RTD structure extends the RTD functionalities to photonic applications.

In its simplest form, a resonant tunnelling diode photodetector (RTD-PD) consists of a DBQW-RTD with relatively thick light sensitive spacer layers adjacent to the DBQW. A vertically illuminated RTD-PD (RTD-VIPD) is a modification of a unipolar semiconductor photoconductive detector, which generically consists of a low-doped/intrinsic semiconductor layer that is sensitive to light sandwiched between high-doped layers to allow ohmic contact to the external circuitry [21–25]. Figure 5.1(a) presents a schematic representation of a double-heterostructure photoconductive detector where the interfaces between the light sensitive regions (here, *n–/i* InGaAs) and the higher bandgap material layers (in this case, InP) used for the contacts form two double heterostructures.

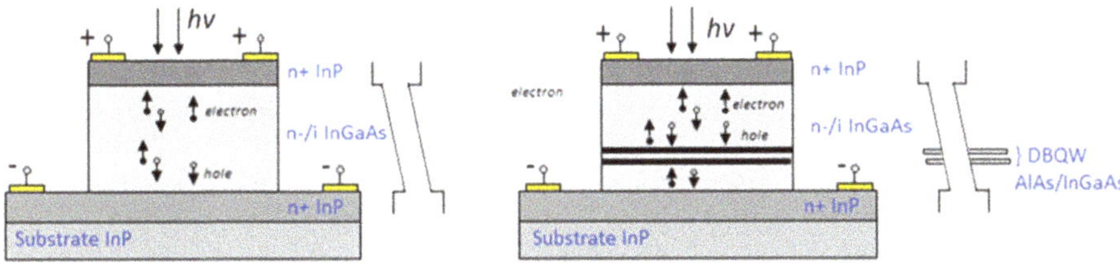

Figure 5.1. (a) Schematic representation of the physical layout of a unipolar InP/InGaAs double-heterostructure photoconductive detector; and (b) of a InP/InGaAs photoconductive detector incorporating an AlAs/InGaAs DBQW consisting of AlAs barriers and a InGaAs quantum well. Sketches of the conduction of the valence band profiles under forward bias are also shown.

A photoconductive detector is a device where the absorption of light increases the number of free charge carriers (electrons and holes) [21]. This raises the overall electrical conductivity of the device. Under bias, the total current flowing in the external circuit is the sum of the dark current (current in the absence of light) and an additional photogenerated current. To a first order, for a given light spectral distribution, the photocurrent intensity is directly proportional to the light irradiance (power per unit area received by the optical window surface).

A double-heterostructure RTD-PD corresponds to the monolithic integration of a DBQW with a unipolar (isotype) *n+/(n–/i)/n+* photoconductive double heterostructure, as schematically shown in figure 5.1(b). The optoelectronic active (light sensitive) region of the device is formed by the low-doped/intrinsic layers and the DBQW. Due to its simplicity, the active region of an RTD-PD is easily monolithically integrated into electronic and photonic devices, adding new functionalities to them.

RTD-PDs are designed to preserve the RTD N-shape current–voltage characteristics and the negative differential conductance (NDC). The photogenerated charge accumulation dynamics enhances the nonlinearities of the electric field-voltage (E-V) profile across the direction of the charge flow, which provides built-in electrical gain (amplification) of the photogenerated signal. Their low-voltage operation, low noise properties, nonlinear photocurrent–voltage and photocurrent–optical power (responsivity) characteristics, and the functionalities offered by the nonlinear N-shape I–V characteristic and the NDC make them alternatives to photodiodes (PDs) and avalanche photodiodes (APDs) [6, 14, 16, 21].

RTD-PDs have been investigated for high-gain high-speed photodetectors, for high-speed optical-to-electrical transducers as needed in future low cost light-wave and radio-over-fibre communication systems, for receivers and spike generators in photonic spiking neural networks (SNNs), and for photon counters and single-photon detectors for quantum-based technologies [4, 5, 14].

In this chapter, we discuss the recent developments in RTD-PDs research. The chapter starts with a review of the RTD-PD concept and its operation, and a brief analysis of the main features of photoconductive detectors that are relevant to the design of high-gain and high-speed RTD-PDs. Next, we present and discuss the main findings of a study of RTD-PD structures. The chapter ends with a discussion of their potential applications as high-speed optical-to-electrical transducers and as light spike generators in photonic spiking neural network (SNN) receivers.

Although in this chapter we focus on RTD-PDs operating in the near-infrared for optical communication and information processing systems, and radio-over-fibre wireless communication systems, the RTD-PD concept has general applicability and has been also considered for mid-infrared detectors [11,12]. In addition, large amplification factors have been reported, with up to 10^5 excess electrons for one absorbed photon [14, 15]. Moreover, several groups are working on RTDs or more general tunnelling devices as integrated amplifiers in photodetectors.

5.2 RTD-PD operation

DBQW RTD-PD heterostructures include at least one relatively thick light-absorbing layer (typically with a thickness between 100 and 500 nm) adjacent to the DBQW. The light-absorbing layers are typically low-doped or intrinsic semiconductor layers, and are, ideally, the only layers that are sensitive to light to be detected. The RTD-PD takes advantage of the RTD's high frequency operation [1], wide-bandwidth electrical gain [3] and low noise [20].

Figure 5.2 shows schematic diagrams of a conventional electronic RTD, panel a), a double-heterostructure RTD-PD configured to work as a vertically illuminated photodetector (RTD-VIPD), panel b), and as an optical waveguide photodetector (RTD-WGPD), panel c). In a conventional electronic RTD, figure 5.2(a), the thickness of the low-doped spacer layers added between the DBQW and the highly-doped emitter and collector contact regions are just few tens of nanometres (the epilayer may include additional undoped layers that are few nanometres thick adjacent to the DBQW that help to suppress diffusion of dopants from the doped layers to the DBQW barriers, preventing dopants from contaminating the barriers).

5.2.1 RTD-PD implementation configurations

The RTD-VIPD for operation in the wavelength range 1200–1600 nm, figure 5.2(b), an InGaAs/InAlAs double heterostructure that confines the InGaAs/AlAs/InGaAs/AlAs/InGaAs DBQW and the intentional added light-absorbing region(s) that is formed by relatively thick (few hundred nanometres) low-doped/intrinsic InGaAs layer(s) (at least ten times thicker than the DBQW). The bandgap energy of the intentionally added layer must be smaller than the energy of the photons to be

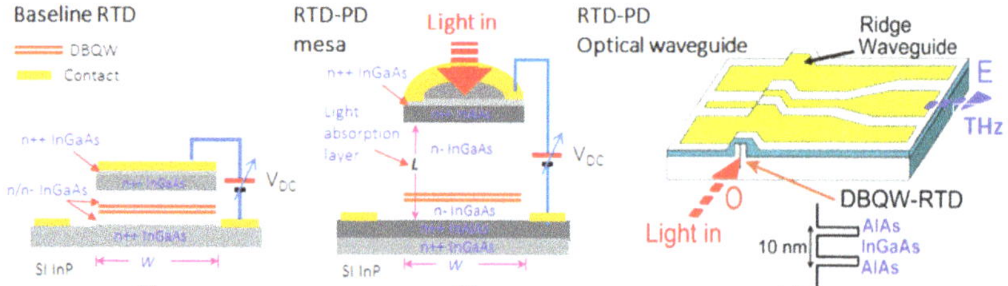

Figure 5.2. Schematic diagrams of an electrical RTD structure (a), and two RTD-PD implementations: (b) mesa/vertically illuminated RTD-PD and optical waveguide RTD-WGPD.

detected. This layer must be embedded together with the DBQW between highly-doped higher bandgap materials.

The RTD-PD in the optical waveguide configuration, figure 5.2(c), has a similar epilayer double heterostructure where the higher bandgap layers act as the cladding layers of the waveguide because they have lower refractive index than the light-absorbing regions. The waveguide core is formed by the DBQW, the adjacent spacer layers plus the intentional added light-absorbing layers. These structures can be implemented using other III–V material compounds, such as GaAs and AlGaAs.

In the RTD-VIPD configuration, figure 5.2(b), an optical window is created on the top contact layer to allow the light to reach the photosensitive regions of the device. The light propagates perpendicularly to the plane of the DBQW/epilayers interfaces. In the RTD-WGPD implementation, figure 5.2(c), the light is butt-coupled to the waveguide with the help of a lens or an optical fibre and propagates in the plane of the DBQW along the light-absorbing layers since these layers have higher refractive index than the surrounding layers confining the light in the plane of the DBQW. The structure of an RTD-WGPD is like the RTD-EAM discussed in the previous chapter.

The highly-doped higher bandgap material layers (the bandgap must be larger than the energy of the photons to be detected) confining the DBQW and the light sensitive layers are used to maximise the photodetector response time (speed of a photodetector). The absorption of the photons in regions of a photodetector where the photogenerated carriers are subject to a low electric field (let's say $\ll 1$ kV cm^{-1}), or no electric field at all, deteriorates its response time. These photogenerated carriers will move by diffusion, which is a slow process when compared with the drifting velocity of carriers under the influence of a large electric field (>1 kV cm^{-1}) [21, 23].

RTD-PD structures should avoid or reduce the absorption of light outside of the high-field regions because the diffusion current can last as long as the carrier's lifetime. The photogenerated current that is being carried by the slow diffusion process leads to the formation of a long tail in the impulse response of the photodiode, which results in a low-frequency falloff in its frequency response. These aspects will be discussed in more detail in the next section.

5.2.2 RTD-PD operation principle

To understand the principle of operation of a RTD-PD, let us first review the RTD's operation principle: as the bias voltage rises from zero, there is a progressive band bending, which is more pronounced in the collector side of the structure, see figure 5.3(c). This gradually lowers the DBQW relative to the emitter spacer, which results in the reduction of the energy difference between the quantised levels in the DBQW and the emitter's Fermi level (Fermi-sea). Consequently, there is an increase in the number of electrons in the emitter that can cross the DBQW through the first available quantised energy level—this process is known as resonant tunnelling. The current through the device rises almost linearly with the applied voltage, building up the first positive differential conductance (PDC) region of RTD-PD I–V characteristic, see figure 5.3(b).

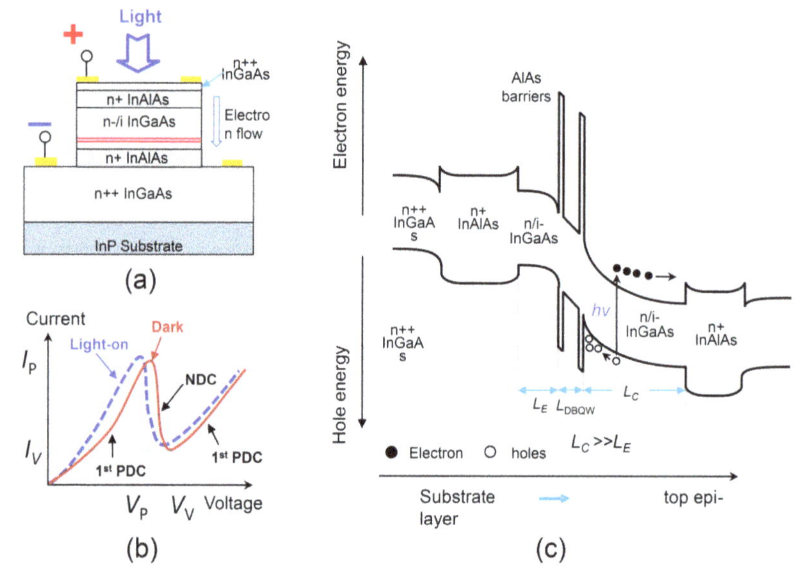

Figure 5.3. (a) Physical layout of an RTD-PD with the light-absorbing region on the collector side. (b) Representation of the DC I–V characteristic under dark and illumination, showing the light induced I–V global shift to lower voltage. (c) Schematic representation of the RTD-PD conduction and valence bands profiles under bias, showing the effects of light absorption: accumulation of holes in the collector region adjacent to the DBQW and electrons drifting towards the collector contact.

In the first PDC region, the current through the device rises with the bias voltage until the voltage reaches a certain value (peak voltage, V_p), which leads to a current local maximum I_p when the DBQW's first available quantised level aligns with the emitter's Fermi level. Increasing the voltage above the peak value leads to a sharp decrease of the current until it reaches a local minimum I_v at the valley voltage value V_v, which happens because there is a reduction of the carrier transmission across the DBQW. As the voltage rises from the peak value, the voltage induced band bending makes the first available quantised level move towards the emitter's band-edge, which progressively reduces the number of electrons that are capable of crossing the DBQW. This current decrease from the peak to the valley corresponds to the device's I–V NDC region. Increasing the voltage above the valley voltage value, other current mechanisms (such as e.g., resonant tunnelling through higher energy levels) make the current increase again. This is the beginning of the second PDC region.

It is important to point out that the voltage drop across the emitter spacer is in general a smaller fraction of the applied bias, and is considerably smaller than the voltage drop across the DBQW and collector spacers. In other words, most of the applied bias is drop across the DBQW and collector spacers, see figure 5.3(c). Consequently, the electric field in the emitter spacers tends to be much weaker than the electric field across the DBQW and the adjacent collector spacers. Another aspect that is worth mentioning is that the filtering effect of the DBQW leads to electron accumulation in the emitter's spacer, which results in the formation of a triangular quantum well adjacent to the DBQW, which impacts on the electric field distribution and on the nonlinearities of I–V in the case of RTDs, and in the photocurrent–voltage and photocurrent–optical power characteristics of RTD-PDs.

The RTD-PD operation principle can be understood similarly (consider again figure 5.3): when a light wave with photon energy higher than the bandgap energy of the light-absorbing layers propagates through the structure, it will be absorbed. This generates electron–hole pairs that modulate the charge transport mechanisms. Besides changing the device's overall conductivity, it also changes the built-in electrical field profile across the structure. The impact on the current–voltage characteristic and on the device photo response will depend on the location of the light-absorbing layers.

Let us first consider the case of forward bias RTD-PD, see figure 5.3(a), where the intentional added light-absorbing layer is in the collector's side, see figure 5.3(c). In the context of RTD polarisation, forward bias means that the top contact layer is at a higher potential than the bottom contact layers and the current follows from the top contact towards to the lower contact. Consequently, electrons move in the opposite direction. The photogenerated electron–hole pairs in the collector's region (where the thickest light-absorbing layer is positioned) are separated by the built-in electric field induced by the applied voltage, giving rise to the photocurrent component.

The photogenerated electrons drift towards the collector's contact with no difficulty. However, the photo-generated holes that drift towards the emitter are blocked by the presence of the DBQW, see figure 5.3(c). These photogenerated holes tend to accumulate in the collector's region adjacent to the DBQW, leading to an electric field enhancement across the DBQW and the adjacent collector depleted region. This means that the resonance condition is achieved at a lower voltage, i.e., under illumination the peak current is reached with a lower bias voltage than when under dark conditions. Therefore, under illumination, RTD-PDs with the light sensitive layers located on the collector's side show a direct current (DC) I–V characteristic overall shift to lower voltage, figure 5.3(b). This overall shift can under certain circumstances come with a significant increase in the current, as it is shown in figure 5.3(b).

The additional photocurrent, which adds to the DC (dark) current of the device, is the result of the enhancement of the device's conductivity under illumination and of the effect of the electron accumulation in the emitter's region on the amount of voltage dropped across the DBQW and its surroundings. This additional photocurrent can become quite significant if the thickness of the emitter spacer layer (which in principle has the same bandgap energy as the collector light sensitive layers) is comparable or larger than the collector's spacer. This aspect will be addressed when discussing the RTD-PD experimental characterisation.

Let us now consider the case of RTD-PDs being reverse biased. As figure 5.4(a) illustrates, now the current flows from the bottom contact to the top contact, and electrons move from top contact layer (emitter) to the bottom contact layer (collector). The light sensitive layer is now acting as the emitter spacer of the device. As shown in figure 5.4(c), the photogeneration of electron–hole pairs happens predominantly in the emitter spacers. Again, the light absorption leads to the enhancement of the device's overall conductivity.

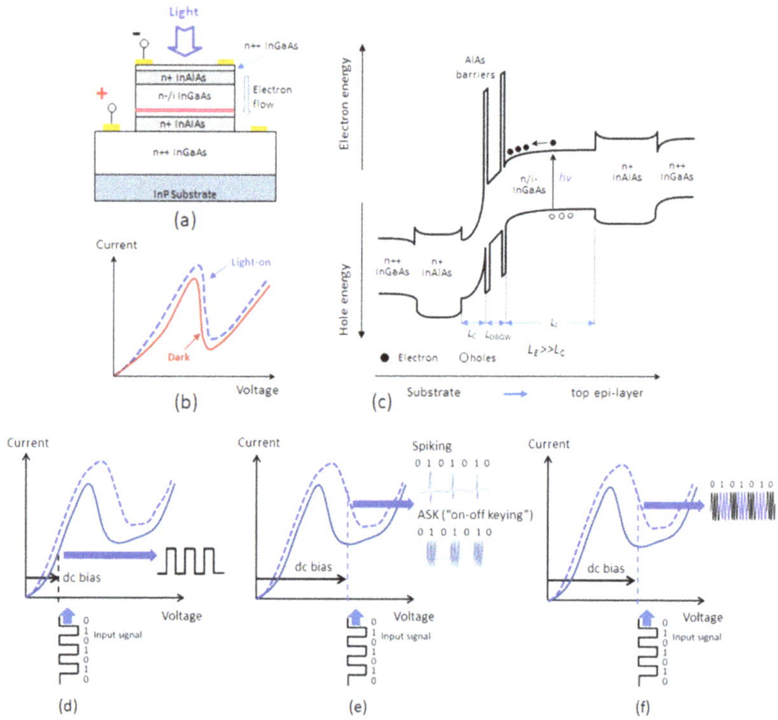

Figure 5.4. (a–c) Schematic representation of the operation of the RTD-PD shown in figure 5.3 under reverse bias. (d–f) Operation modes of an RTD-PD circuit. (d) Non-oscillating mode: device biased in the PDC sections of the *I–V* curve. (e) Excitable mode and optical induced electronic oscillator (amplitude shift keying (ASK)—'on-off keying'): devices are biased close to but outside the NDC region (here in the second PDC region). (f) Oscillation mode (frequency shift keying (FSK)): device biased on the NDC. The dashed lines represent the *I–V*s under illumination.

As mentioned earlier, the photogenerated electron and hole pairs in the emitter spacers are in general subject to much weaker electrical fields than when they are photogenerated in the collector spacers. Under the influence of this weaker electric field, the photogenerated holes tend to move towards the emitter's high-doped contact (the top layer) and the photogenerated electrons travel in the direction of the high-doped collector's contact (the bottom layer). However, the flow of the photogenerated electrons is constrained by the filtering effect of the DBQW, which tends to reinforce the electron accumulation in the emitter region adjacent to the DBQW.

When these effects are combined they reduce the voltage dropped in the emitter and make the effect of the applied voltage on the band bending of the DBQW less effective. So, as a consequence of the illumination, the voltage required to achieve the resonance (peak current) is larger than under dark conditions. This, together with the conductivity enhancement, leads to an overall shift of the *I–V* to higher voltage, which is accompanied by a current increase.

Most RTD-PDs mentioned in the literature have light sensitive layers in the collector's region adjacent to the DBQW. There is a consensus in the RTD-PD community that the I–V shift to lower voltage, as shown in figure 5.3(b), is due to the enhancement of the built-in electrical field across the DBQW and the adjacent depleted collector region as a consequence of the accumulation of the photo-generated holes, as represented in figure 5.3(b). Few groups have reported RTD-PDs showing I–V overall shift to high voltage, see figures 5.4(a)–(c). These devices will be discussed in more detail in the section dedicated to the experimental evaluation of RTD-PDs epilayers.

The RTD-PD operation takes advantage of nonlinearities in the electrical field profile in the direction of the current flow. These electrical field nonlinearities are induced by the carrier filtering effect of the DBQW, and by charge accumulation in the emitter and collector spacers adjacent to the DBQW. Under certain operation conditions, these nonlinearities provide electrical amplification for the photocurrent, giving rise to a significant enhancement of the RTD-PD responsivity, which is strongly dependent on irradiance and applied voltage.

5.2.3 RTD-PD modes of operation

Generically speaking, high-frequency RTD-PD circuits can operate in four modes, as shown figures 5.4(d)–(f): (i) non-oscillating (steady state/passive mode); (ii) optical induced excitability; (iii) optical induced oscillations (optical induced amplitude shift keying (ASK)—'on-off keying'); and (iv) self-oscillation (optical induced frequency shift keying (FSK)). In what follows, we will consider devices under illumination that produce an I–V overall shift to higher voltage. The operation conditions for each mode are determined essentially by the electrical circuit parameters, the RTD-PD I–V characteristics and the characteristic of the optical signals. In all these operation modes, the optical-to-electrical conversion benefits from the built-in gain due to the device's nonlinearities.

In the 'passive mode', the devices are in general biased in the first PDC region sufficiently far from the NDC, figure 5.4(d). In this mode, the RTD-PD shows similar photo response to PIN-PDs and APDs, converting the intensity modulated optical signal into an electrical signal that resembles the electrical signal used to modulate the light-wave signal.

The optically induced excitability mode can be observed when the steady state quiescent point of the RTD-PD circuit is in the second PDC region but close to the NDC, see figure 5.4(e). To run in the excitable regime, the circuit must be designed to operate as a monostable system. (A monostable system is a system that has one stable state and one quasi-stable state.) A sufficient large intensity light pulse induces the RTD-PD operation point to switch from the second PDC to the first PDC, returning to the starting point, following a behaviour known as an all-or-nothing response. Under these operating conditions, whenever the intensity of a modulated optical signal reaches a given threshold, the circuit reacts by producing current and voltage pulses (spikes), and in some cases bursts of spikes. This mode of operation will be discussed in more detail later in this chapter.

When not designed to work as a monostable circuit, the RTD-PD circuit can operate as an optically induced electronic oscillator whenever the intensity of a modulated optical signal is sufficient to move its quiescent operation point to the NDC region. The circuit produces current and voltage oscillations lasting the time of the duration of the high intensity portion of the optical signal, in a process that is identical to the modulation scheme known as ASK or 'on-off keying'[26], see figure 5.4(e). Under these conditions, the RTD-PDs can operate as optical-to-microwave transceivers and optically controlled RTD oscillators. Additionally, when the modulation frequency of optical signal is in the range of the RTD-PD self-oscillation frequency, the circuit can produce current changes with amplitude equivalent to the peak-to-valley current difference. This can result in absolute responsivities above hundreds of ampere/watt (A W^{-1}).

Whenever the devices under dark conditions are biased in the NDC region, see figure 5.4(f), they can be in a self-oscillation mode. A modulated light signal can either produce FSK modulation of the self-oscillation signal or the turning off the self-oscillations [13, 26, 36]. Under certain conditions, such as the injection of relatively low-level optical signals, the circuit can be operated in the injecting locking mode: the modulated optical signal locks the circuit oscillations to the modulation frequency of the optical signal, a process known as optical injection locking (which will be discussed in more detail in the following chapters).

As mentioned, RTD structures incorporating light sensitive layers can be operated as high-frequency electronic oscillators that are capable of being controlled by optical signals. This functionality has noticeable applications in signal processing and telecommunications. Due to their wide-bandwidth NDC, RTD-PDs offer several modes of operation that can add new functionalities when hybridly combined or monolithically integrated with other electronic and optoelectronic devices. Among the functionalities to be considered, with applications in telecommunication data links as optical-to-wireless converters and as local oscillators, it is worth mentioning their operation as voltage controlled optoelectronic oscillators and as optical controlled microwave oscillators.

Another interesting feature of RTD-PD devices is that when the bias voltage is set in the PDC regions but not far from the NDC region, the injection of a high-frequency pulsed optical signal can induce fast changes in the conductivity of the device, giving rise to a high-frequency pulsed current flowing in the external circuit. This pulsed current can be used to directly drive other electronic and optoelectronic components, such as semiconductor lasers [5, 27]. This capability is being used in novel optoelectronic systems such as optoelectronic oscillators [28], optical/wireless transducers links [29] and neuromorphic optoelectronic circuits [18].

Compared with PIN-PDs and APDs [21–23, 30–32], RTD-PDs have the potential to achieve much higher bandwidth-efficiency products due to their high-speed, considerably superior value of effective responsivity and their high nonlinear photocurrent–irradiance characteristic [7–17]. Moreover, the RTD-PD electrical amplification mechanism is, in principle, a much lower noise process when compared with the APD carrier multiplication mechanism through impact ionisation. The RTD-PD electrical gain noise is not subject to random impact ionisation processes but is subject to general thermal considerations. However, it is expected that the RTD-PD bandwidth will be limited by parasitic components, as is the case with the RTD oscillators which have been demonstrated to operate up to 1.92 THz [1], it is expected they can operate well into the GHz range. Overall, when compared to the APD and PIN-PD, the RTD-PD structures can operate at much lower voltages (typically around or below 1 V), have potentially lower noise levels and have a superior bandwidth-efficiency product.

5.3 Review of photoconductive detectors

A photodetector is an optoelectronic light sensor where the absorption of a light signal induces changes in the device's characteristics, e.g., its electrical conductivity, that are proportional to the light properties, e.g., the light intensity [21]. Here the discussion is centred on photodetectors based on the photoconductive effect, i.e., photoconductive detectors. In photoconductive photodetectors, light absorption leads to the generation of electron–hole pairs in the material, changing its electrical conductivity, i.e., photoconductive detectors convert a photon flux into an electrical current. These photodetectors are typically made of semiconductor materials [21, 24, 25].

The main requirements for photodetectors include large responsivity (defined as the ratio of the photocurrent and the optical power incident on the photodetector illumination window), high sensitivity (defined as the minimum amount of optical power incident on the photodetector that can produce a distinguished signal, i.e., when the signal-to-noise ratio is equal to 1), low noise, wide optical bandwidth, high-speed response, high reliability and low cost [21–32].

5.3.1 Operating principle of photoconductive detectors

The operating principle of most common semiconductor-based photodetectors can be summarised as follows [21]: if photons incident on their illumination window have an energy larger than the bandgap energy of a given layer of the semiconductor stack, then the light will be absorbed in that particular layer, creating electron–hole pairs. This increases the semiconductor's electrical conductivity and thereby changes the intensity of current flowing in an external circuit, which can then be converted into a voltage signal. The magnitude of the photogenerated current or voltage is proportional to the number of incident photons being absorbed.

Most common semiconductor photodetectors employ a light sensitive region that corresponds to the depletion region of a *PN* junction, or to a low-doped or intrinsic (I) semiconductor layer embedded within a *PN* junction forming a *PIN* hetero-junction, see figure 5.5. These structures are known as photodiodes (PDs) when operating under reverse bias, i.e., the P side is made more negative than the N-side. The reverse bias reinforces the built-in electric field in the depletion or in the intrinsic region of the PN/PIN diode, which separates and drifts the photogenerated electrons and holes, producing an additional current in the external circuit that is proportional to the light intensity.

A photodiode behaves as a photo-controlled current source in parallel with a semiconductor diode, governed by the standard diode equation, both in parallel with

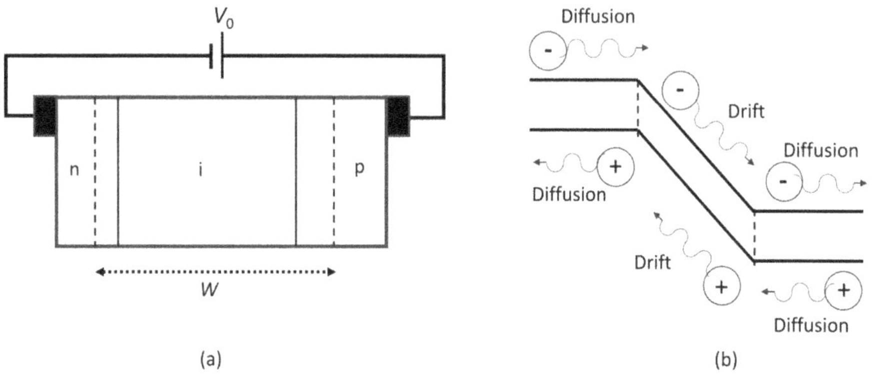

Figure 5.5. (a) The structure of a PIN photodiode. (b) Energy-band diagram showing carrier movement through drift and diffusion: the electron–hole pairs generated inside the depletion and intrinsic (intermediate) region experience a large electric field and drift rapidly toward the *p*- and *n*-side, depending on the electric charge polarity. The resulting current flow constitutes the photodiode response to the incident optical power.

the detector shunt resistance, i.e., the slope (dV/dI) of the diode I–V curve evaluated at zero bias. The total current flowing through the device is the sum of the diode current (known in the photodetection context as dark current) and the photo-generated current. When operating under reverse bias, the dark current is simply the diode's reverse leakage current.

There are two main modes of operation of the photodiode [21]: photoconductive and photovoltaic. In the case of the photoconductive mode, an external reverse voltage is applied to create a high-electric field in the light sensitive regions, which makes the photogenerated charges move away at high speed from the absorption region to the photodetector contacts. This mode of operation is required for high response speed. The reverse bias current is mainly due to minority photogenerated carriers. The diffusion current due to the photogenerated electron–hole (e–h) pairs outside the depletion and intrinsic regions tends to decrease the speed of response because diffusion is a slow process.

In the photovoltaic mode, no voltage or only a small direct voltage is applied and the photogenerated current flows through the detector shunt resistance, causing a voltage across the diode. This voltage opposes the barrier voltage of the photodiode junction, forward biasing the diode. The value of the detector's shunt resistance drops exponentially as the illumination increases. Thus, the photogenerated voltage is a logarithmic function of incident light intensity.

In addition to this basic photodetection mechanism, some photodetectors can include an electrical amplification section, such as in Avalanche Photodiodes (APDs) [23]. APDs are advanced PIN-PD structures that include extra semi-conductor layers to provide internal gain as a result of the carrier multiplication process through (for example) impact ionisation, which increases the detection sensitivity with a relatively low noise penalty [22, 23, 30, 32].

In high-speed optical communications, PIN-PD and APD are operated in the photoconductive region well away from the photovoltaic region, i.e., they are

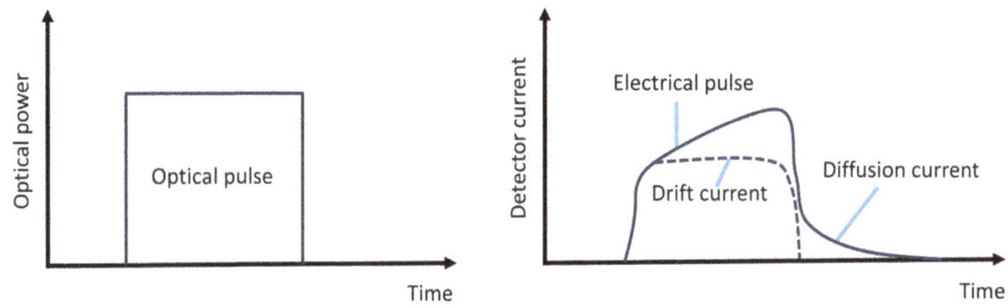

Figure 5.6. Response of a photodiode to a rectangular optical pulse when both drift and diffusion currents contribute to the detector's overall photocurrent.

reversed bias (the cathode is made positive with respect to the anode). In the case of PIN-PDs, the reverse voltage is typically in the 5–20 V range. The bias voltage of APDs can reach several hundred volts.

The limiting factor for the bandwidth (response time) of PN/PIN photodiodes is the presence of a diffusion component in the photocurrent. The physical origin of the diffusive component is the absorption of incident light outside the depletion region and/or the intermediate intrinsic semiconductor layer (see figure 5.5). The e–h pairs generated in the P and N regions (low electric field regions) must first diffuse into the depletion and intrinsic regions before being swept away to the highly-doped contacts. Electrons generated in the p-region must diffuse to the depletion/intrinsic-region boundary before they can drift to the *n*-side. Similarly, holes generated in the *n*-region must first diffuse to the depletion-intrinsic region. Carrier diffusion is a slow transport process: the carriers take typically a nanosecond or longer to diffuse across 1 μm. Figure 5.6 schematically shows how the presence of a diffusion current component can distort the temporal response of a photodiode. Additionally, these e–h pairs may go through recombination before reaching the contacts, resulting in a reduced current, which lowers the responsivity.

The performance of PIN photodiodes can be improved using double-heterostructure designs: an intrinsic/low-doped layer is sandwiched between P-type and N-type layers, which are now made of semiconductor compounds with a high enough bandgap energy to not absorb light. Now, the photons are only absorbed in the lower bandgap intrinsic layer. A PIN photodiode that is commonly used in the wavelength range 1200–1600 nm employs $In_{1-x}Ga_xAs$ lattice matched to InP (with $x = 0.47$) for the intrinsic layer and InP for the surrounding P-type and N-type layers (forming two InP/InGaAs double heterostructures). Since the bandgap of InP is 1.35 eV, InP is transparent for light whose wavelength exceeds 0.92 μm. The bandgap of $In_{0.53}Ga_{0.47}As$ (lattice matched to InP) is about 0.75 eV, a value that corresponds to a cutoff wavelength of 1.65 μm.

Figure 5.7 shows the absorption coefficient spectra $\alpha(\lambda)$ for the major semiconductor materials or compounds employed to implement photodetectors as a function of light wavelength λ, including the InGaAs ternary and InGaAlAs quaternary compounds lattice matched to InP. InGaAs is widely used for photodetection in the wavelength range 1.2–1.6 μm because it has a quite significant absorption coefficient value.

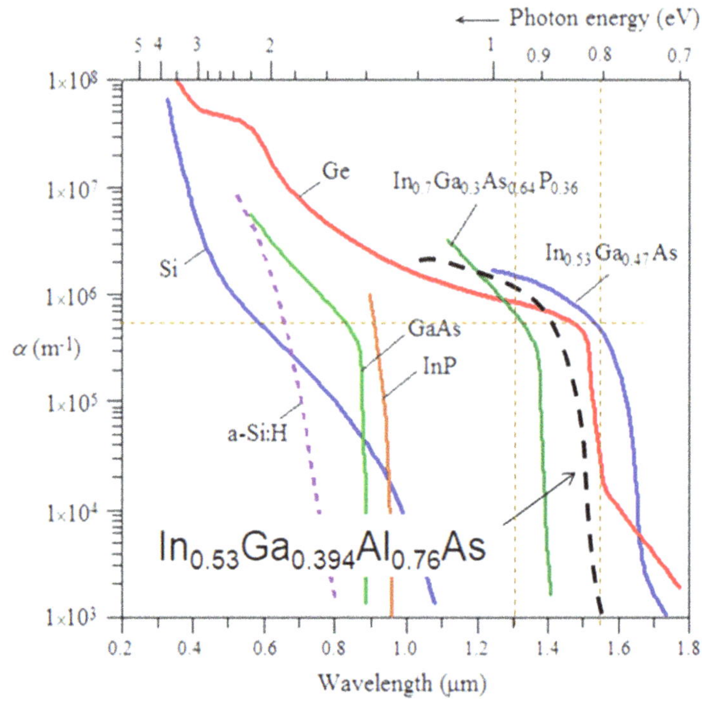

Figure 5.7. Absorption coefficient $\alpha(\lambda)$ as a function of the wavelength λ for several semiconductor materials [21, 33].

Using these InP/InGaAs double heterostructures eliminates the diffusion component of the detector current due to the photogenerated e–h pairs in the high-doped regions because photons are absorbed only in the low bandgap depleted regions and in the intrinsic or low-doped low bandgap layers. To improve the responsivity, the photodetector's optical window can be coated with an antireflection coating to minimise reflections (insertion losses) at the semiconductor–air interface [21].

5.3.2 Design guidelines of high-performance photodetectors

The efficiency of a photodetector as an optical-electrical converter is normally characterised by its responsivity, which is defined as the ratio of generated photocurrent and the incident optical power. In an ideal photodetector, every photon produces one electron–hole pair that is transferred instantly and without losses to the contact electrodes of the photodiode, giving rise to the light-generated current [25], i.e., the quantum efficiency of an ideal photodetector is equal to one.

Optical-based signal processing and communication systems require photodetectors with responsivity as large as possible, high-speed response, high sensitivity and low noise [25]. The performance of the recovering process of the information carrying signal is highly dependent on the noise level at the receiver. Additionally, in general, the electrical output of a photodetector requires an amplification stage [25].

Consequently, it is of paramount importance to have high-bandwidth photodetectors with high responsivity, high sensitivity and low noise.

The RTD-PD characterisation employs the main photodetection standard metrics (it is worth pointing out that these photodetection standard metrics are defined for linear systems) [25]. This includes responsivity (which is defined as the ratio of generated photocurrent and the incident (and absorbed) optical power, determined in the linear response region [21, 24, 25]), sensitivity (the smallest photosignal that can be detected related to the noise or noise equivalent power) and the photodetection bandwidth (speed of response or response time).

The responsivity of a photodetector is usually defined for the steady state [25]. The term *responsivity* should not be confused with *sensitivity*—the latter is the lowest detectable light level, which is typically determined by the noise level of the receiver and is significantly influenced by the required detection bandwidth [25]. For a PIN-PD, the responsivity, $\mathcal{R}(\lambda, V)$, in the linear part of its light-current characteristic is given by [24]:

$$\mathcal{R}(\lambda, V) = \frac{I_{ph}(\lambda, V) - I_{\text{dark}}(V)}{P_{\text{ph}}(\lambda)} \approx \frac{I_{\text{ph}}(\lambda, V)}{P_{\text{ph}}(\lambda)} = \eta(\lambda, V)\frac{e\lambda}{hc} \tag{5.1}$$

where $I_{\text{ph}}(\lambda, V)$ is the photogenerated current, which is typically much larger than the dark current ($I_{\text{ph}}(\lambda, V) \gg I_{\text{dark}}(V)$); $\eta(\lambda, V)$ is the quantum efficiency, which is a function of the optical material proprieties (absorption coefficient $\alpha(\lambda, V)$, refractive index $n(\lambda, V)$), the thickness L of the absorbing region, and operating voltage V); and e, h, and c are the elementary charge, Planck's constant and the speed of light in vacuum, respectively.

Considering a photodetector with the maximum quantum efficiency, i.e., with $\eta = 1$, equation 5.1 predicts that the responsivity, $\mathcal{R} = \eta(\lambda, V)\frac{\lambda(\mu\text{m})}{1.24}$, for light with wavelength about 1240 nm will be 1 A W^{-1}. Assuming that the quantum efficiency η does not change with wavelength, the responsivity will be roughly 1.25 A W^{-1} for light with wavelength of 1550 nm.

The effective responsivity of a photodetector can be increased by incorporating electrical gain [21]. However, the electrical gain must be integrated without degrading the photodetector's optimal performance, such as introducing the minimum noise and not reducing its bandwidth. For APDs, the electrical gain noise (or so-called excess noise) arises from the random nature of impact ionisation. The bandwidth of the APD electrical gain is determined by the avalanche build-up time and the so-called dead time over which the avalanche build-up take places, and both are subject to random processes [31]. For photodetector devices with electrical gain, G, such as APDs, the effective responsivity is as follows [21–26]

$$\mathcal{R}_G(\lambda) = G(V, P_{\text{ph}})\eta(\lambda)\frac{e\lambda}{\text{hc}} = G(V, P_{\text{ph}})\eta(\lambda)\frac{\lambda(\mu\text{m})}{1.24} \tag{5.2}$$

where $G(V, P_{\text{ph}})$ is the average number of electrons flowing in the external circuit per photogenerated electron–hole pair.

To maximise the sensitivity of the device, i.e., to reduce the amount of optical power needed to produce a detectable response in the photodetector, the dark current must be made as low as possible. In PIN-PDs, the dark current can be made very small (below 1 nA) and the total current following through the external circuit is in practice directly proportional to the light irradiance [31]. When APDs are employed to provide internal gain, it is crucial to keep the noise level as low as possible because the performance of the signal recovering process is highly dependent on the noise level at the receiver [30, 31].

Photodetector standard metrics are defined for linear systems and must be used with care when characterising and designing receiver systems that employ RTD-PDs. In particular, the responsivity $\mathcal{R}$ must account for the fact that RTD-PD I–V characteristic and the electric field profile along the direction of propagation of the light are highly nonlinear, making RTD-PD a highly nonlinear optoelectronic device. Consequently, RTD-PDs, in general, do not have a single value for $\mathcal{R}(\lambda)$ that covers the full voltage range of its operation for a given wavelength λ. It is expected that the responsivity of RTD-PDs will be dependent on the bias voltage V and on the power of the optical signal and its wavelength λ, i.e., $\mathcal{R}_{\mathrm{RTD-PD}} = \mathcal{R}(\lambda, V, P_{\mathrm{ph}})$.

The RTD-PD responsivity in the steady state (or stationary mode) corresponds to the photodetection response to an incident continuous light wave, i.e., a constant intensity optical signal. The RTD-PD steady (stationary) state responsivity $\mathcal{R}_{\mathrm{RTD-PD}} = \mathcal{R}(\lambda, V, P_{\mathrm{ph}})$, measured in A W^{-1}, is calculated by:

$$\mathcal{R}_{\mathrm{RTD-PD}} = \left| \frac{I_{ph}(\lambda, V, P_{\mathrm{ph}}) - I_{\mathrm{dark}}(V)}{P_{\mathrm{ph}}(\lambda)} \right| = G(V, P_{\mathrm{ph}})\eta(\lambda)\frac{e\lambda}{\mathrm{hc}} \tag{5.3}$$

where $I_{\mathrm{ph}}(\lambda, V, P_{\mathrm{ph}}) - I_{\mathrm{dark}}(V)$ is the current difference under illumination and dark conditions; $P_{\mathrm{ph}}(\lambda)$ is the continuous-wave optical power incident on the device; and $G(V, P_{\mathrm{ph}})$ accounts for the internal current gain effect (due to electron–hole pair generation, built-in electrical gain, and/or impact ionisation mechanisms, or other phenomena not yet clearly identified).

We can estimate the external quantum efficiency and the responsivity of RTD-PDs in the vertically illuminated and in the waveguide configurations. The external quantum efficiency $\eta(\lambda)$ can be estimated using the relation:

$$\eta(\lambda) = \kappa(1 - R_{\mathrm{ref}})(1 - e^{-\alpha(\lambda)L}) \tag{5.4}$$

where κ accounts for the internal quantum efficiency (how many electron–hole pair are generated per photon absorbed within the light sensitive layers), which we assume to be $\kappa = 1$ for the time being; R_{ref} is semiconductor reflectance, $R_{\mathrm{ref}} = \left(\frac{n-1}{n+1}\right)^2$, with $n(\lambda)$ being the semiconductor refractive index. The term $(1 - R_{\mathrm{ref}})$ can be made $(1 - R_{\mathrm{ref}}) \approx 1$ using an illumination window with an antireflection coating. Since the responsivity is directly proportional to the external quantum efficiency, it can be estimated by:

$$\mathcal{R}_G(\lambda, \mathrm{V}) = G(V, P_{\mathrm{ph}})\frac{e\lambda}{\mathrm{hc}}\kappa(1 - R_{\mathrm{ref}})(1 - e^{-\alpha(\lambda)L}) \tag{5.5}$$

An absorption layer with a thickness L larger than the reciprocal of the absorption coefficient (i.e. $L > 1/\alpha(\lambda)$) is needed to achieve high quantum efficiency, and hence high responsivity. However, a large absorbing layer will result in a longer carrier drift time (and eventually diffusion time), which is detrimental to the speed response of the device. This can be overcome using a waveguide photodetector configuration: the absorption layer along the direction of the light propagation can be made simultaneously much larger than $1/\alpha(\lambda)$ to maximise the responsivity and sufficiently thin to minimise the carrier's drift time.

Several effects and loss mechanisms can lead to a reduction of responsivity. In the case of vertically illuminated configuration (as mentioned previously), the photodetection quantum efficiency value is determined by the thickness of the light absorption layers, by the optical reflection at the semiconductor–air interface (which can compass for 30% of the incident light), the losses in the light coupling (e.g. the case of optical fibre-chip coupling), the losses in the internal contact layers and by carrier recombination, which may occur before the carriers reach the contacts.

For high-speed digital optical communication systems and many other applications, it is required that the individual components, such as photodetectors, have a frequency response $H(\omega)$ with a 3 dB frequency (or 3 dB bandwidth), which is well above the corresponding maximum foreseen data bit rate for the systems [21–26]. The 3dB bandwidth is the frequency at which the magnitude of the response (current or voltage) has fallen off by 3 dB, i.e., the frequency at which the current or voltage reach 70% of the low-frequency value.

For optical telecommunications, the bandwidth metric is crucial because it largely determines the data rates, which are now in the range 10–100 Gb s^{-1} [26]. In a single channel telecommunication system, the overall bandwidth is limited by the bandwidth of the smallest bandwidth component. In the case of an optical receiver, the bandwidth is limited by the physical processes in the photodetector and by how the parasitic inductances, resistances and capacitances contribute to the time constants of the external circuit

As mentioned earlier, the photodetector response typically falls off for intensity modulated signal frequencies above some detection frequency. There are basically three factors that limit the speed of a photodetector [21, 34]: the diffusion of the carriers, the drift transit time in the depletion region and the capacitance of the depletion region. The slowest of the three processes is the diffusion of carriers to the high-electric field depletion region from outside that region. To minimise this slow effect, carriers should be generated near or in the depletion region. Reducing the ratio between the diffusion length and the drift length of the device results in a greater proportion of the generated current being carried by the faster drift process.

In what follows, we assume that the critical distances or thicknesses (e.g. light-absorbing layer thickness, the depleted and intrinsic/low-doped layer thicknesses and the region where the diffusion takes a role) have a maximum magnitude equal to 500 nm. We are aware that these assumptions may be unrealistic because they do not consider relevant aspects, e.g., the depletion regions are bias dependent, while the thickness of the light sensitive layers is fixed by the epilayer design.

Diffusion of the carriers generated outside the depletion region is a relatively slow process, which is caused by the optical absorption in the volumes outside of the high-electric field regions. The diffusion component of the current can last as long as the carrier's lifetime, and they tend to generate a long tail in the impulse response of the photodetector, which results in a low-frequency falloff in the frequency response. The diffusion time τ_{diff} for carriers to diffuse a distance Λ can be estimated using the following equation

$$\tau_{\text{diff}} = \frac{\Lambda^2}{D} \tag{5.6}$$

where D is the minority carrier-diffusion coefficient. For InGaAs electron and hole diffusion, the coefficients are of the order of 150 and 5.5 $\text{cm}^2\ \text{s}^{-1}$ [21], respectively, which, for a diffusion distance of $\Lambda = 500$ nm, gives a hole diffusion time τ_{diff} around 227 ps.

The 3 dB electrical frequency (i.e. frequency where the collected current has decreased by 3 dB) of a photoconductive detector limited by the carrier-diffusion time can be estimated using the following equation [21, 26]:

$$f_{3\text{dB}-\text{diff}} \approx \frac{3.5}{2\pi}\frac{1}{\tau_{\text{diff}}} \tag{5.7}$$

Using the value of $\tau_{\text{diff}} = 227$ ps gives a 3 dB frequency $f_{3\text{dB}-\text{diff}} \sim 3.3$ GHz.

For a high-speed photodetector, this diffusion mechanism must be eliminated or highly reduced, e.g., by reducing the photogeneration of carriers outside the depletion layer through an improved design of the device structure such as incorporating a double heterojunction.

The second process, the transit time, is the time required for the carriers to drift across the depletion and the intrinsic regions and get swept out of the device. For higher frequency operation, the transit time of the photogenerated carriers, particularly the minority carriers, across the structure must be small, which implies that the length L of the depletion and intrinsic regions must be minimised. Additionally, the photogenerated carriers' velocities must be maximised. At sufficiently high reverse bias voltage, electrons and holes will drift at their saturation velocities [21, 26]. The 3 dB electrical frequency of a photoconductive detector limited by the carrier-transit time can be calculated using [21, 26]:

$$f_{3\text{dB}-\text{tt}} \approx \frac{3.5v}{2\pi L} \tag{5.8}$$

with $v^{-4} = \frac{1}{2}v_e^{-4} + \frac{1}{2}v_h^{-4}$, where v_e and v_h are the electron and hole saturation velocities, respectively. For very high-bandwidth photodetectors, the bandwidth-efficiency product $f_{3\text{dB}-\text{tt}} \times \eta_{\text{ph}}$ [21, 26]:-

$$f_{3\text{dB}-\text{tt}} \times \eta_{\text{ph}} \approx \frac{3.5v}{2\pi L}\kappa(1 - R_{\text{ref}})(1 - e^{-\alpha(\lambda)L}) \tag{5.9}$$

The expected intrinsic carrier-transit time limited 3 dB frequency f_{3dB-tt} of a photodetector with a light-absorbing length of L = 500 nm is of the order of 60 GHz, assuming $v = 5.3 \times 10^6$ cm s^{-1} (from electron and hole velocities in InGaAs, $v_e = 6.5 \times 10^6$ and $v_h = 4.8 \times 10^6$ cm s^{-1}, respectively [26]). This corresponds to a transit time of around 10 ps.

For vertically illuminated photodetectors, i.e., structures where the illumination is perpendicular to the epilayer interfaces, several compromises in designing must be considered with respect to the thickness of the photon-absorbing layer. The bandwidth-quantum efficiency product for vertically illuminated photodetectors depends on the operating wavelength. A large absorption layer length is needed to achieve high quantum efficiency, and hence high responsivity. However, a long absorbing layer results in a large carrier drift time. As a numerical reference and assuming single-pass absorption, it is estimated that for a material having a light-absorbing coefficient of the order of 1 μm^{-1}, the ideal bandwidth-efficiency product in the third and second optical communication windows will be below 20 and 35 GHz, respectively [26].

Finally, the capacitance of the photodetector determines its RC-time constant (here, R represents the device/circuit load resistance, usually 50 Ω) [26]. To maximise a photodiode's response, the transit time and diffusion time are typically designed to be comparable but smaller than the RC-time constant. For instance, given the saturation velocity for InGaAs, a 10 ps transit time (for example) requires that the depletion layer to be no longer than 0.5 μm at most. For a comparable RC-time constant in a 50 Ω system, the capacitance must be less than 200 fF. Assuming that the capacitance C can be estimated using the parallel plate approximation,

$$C \approx \epsilon \epsilon o \frac{A}{d} \tag{5.10}$$

where d is the dielectric thickness (here the depleted layers plus the intrinsic layer thicknesses), A is the active area of the device and ϵ is the relative material permittivity ($\epsilon \approx 13$ for InGaAs), and ϵo is the vacuum permittivity.

Assuming d = 0.5 μm, and considering the numerical values above, we find that the photodetector active area A must be less than 870 μm^2 (equivalent to disc with a diameter smaller than 35 μm).

For photoconductors [24], the recombination times, τ, of electrons and holes dominate many aspects of the design of large bandwidth devices. Reverse biased PIN photodiodes tend to be faster than photoconductors because they have a wide intrinsic section inserted between the P and N regions, which allows them to attain very large built-in fields. This separates and rapidly sweeps away the electrons and holes towards the electrodes. This wide intrinsic section leads to lower capacitance, and thereby to a smaller RC-time constant.

Optical telecommunication photodetectors are mainly optimised with respect to high bandwidth-quantum efficiency product (or gain bandwidth product), sensitivity (lowest detectable light level) and high-output saturation power (responsivity) [26].

5.4 RTD-PD implementation and characterisation

In this section, we present the main findings on the performance characterisation of several InGaAs/InP RTD-PD epilayers operating as vertically illuminated photodetectors. The characterisation is focused on the RTD-PDs' responsivity and how it compares to that of the PIN-PDs and APDs. In addition, we will discuss the preliminary results of the response time of these vertically illuminated RTD-PDs.

This investigation of the InGaAlAs/InP RTD-PD epilayer structures aims to examine the potential of this new class of photodetector to take advantage of the RTD nonlinearities and, in particular, of the NDC region associated with the DBQW-RTD structures.

5.4.1 RTD-PD responsivity

Table 5.1 describes in detail an InGaAs/AlAs RTD-PD epilayer structure comprising a 500 nm low-doped InGaAs layer for light absorption within the wavelength range 1200–1600 nm. Notice that, excluding the AlAs barriers, all of the epilayers are made of InGaAs lattice matched to InP, i.e., the devices lack the two double heterostructures mentioned in the previous sections.

Figure 5.8 shows the physical layout of InGaAs/InP RTD-PD, and the epilayer conduction and valence band profiles. Mesa RTD-PD devices 10 μm × 10 μm with 5 μm × 5 μm optical windows were implemented. A single mode optical fibre was used to couple light through the optical window defined on the top of the mesa.

The devices showed peak current densities around 0.3 mA μm^{-2} and peak-to-valley current ratios close to 1.5. Under dark conditions, the devices' NDC region was between 1.1 and 1.4 V. The stationary photodetection responses were

Table 5.1. Epi-layer structure of an InGaAs/InP RTD-PD designed to operate in the wavelength range of 1200–1600 nm. Excluding the AlAs barriers, all of the InGaAs epilayers absorb light in that wavelength range. The layers' description assumes that the devices are forward biased.

Layer #	Thickness (nm)	Material	Doping (cm^{-3})	Description
1	45	$In_{0.53}Ga_{0.47}As$	3 × 1019: Si	Collector
2	80	$In_{0.53}Ga_{0.47}As$	3×10^{18}: Si	Collector
3	500	$In_{0.53}Ga_{0.47}As$	5×10^{16}: Si	Absorption
4	2	$In_{0.53}Ga_{0.47}As$	Undoped	Spacer
5	1.7	AIAs	Undoped	Barrier
6	4.7	$In_{0.53}Ga_{0.47}As$	Undoped	Well
7	1.7	AIAs	Undoped	Barrier
8	2	$In_{0.53}Ga_{0.47}As$	Undoped	Spacer
9	20	$In_{0.53}Ga_{0.47}As$	2×10^{16}: Si	Spacer
10	80	$In_{0.53}Ga_{0.47}As$	3×10^{18}: Si	Emitter
11	400	$In_{0.53}Ga_{0.47}As$	3×10^{19}: Si	Emitter
12	200	InP	Undoped	Buffer
		InP substrate		

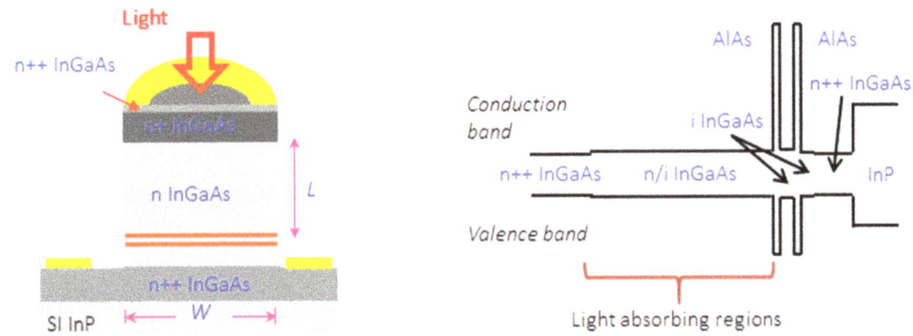

Figure 5.8. Physical layout of InGaAs/InP RTD-PD mesas with a light-absorbing layer 500 nm thick. Also shown is a representation of the device conduction and valence band profiles at zero bias.

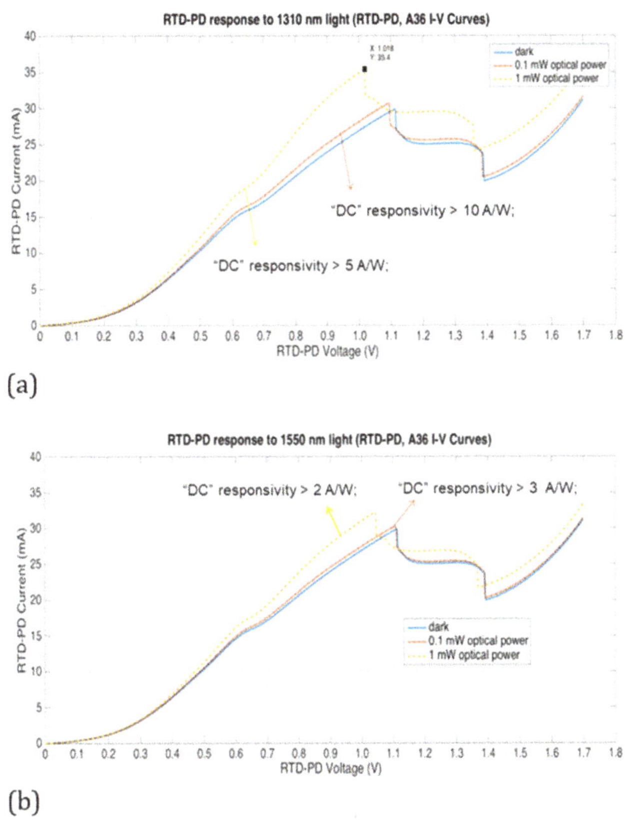

Figure 5.9. Measured current–voltage response of mesa InGaAs/InP RTD-PD at wavelengths (a) 1310 nm and (b) 1550 nm.

characterised using unmodulated light with wavelengths of 1310 and 1550 nm, respectively, for three optical power levels: dark, 0.1 and 1 mW. Figure 5.9 shows the typical I–V characteristics under dark conditions and for illumination (a) at 1310 nm and (b) at 1550 nm.

It is clearly noticeable in figure 5.9 that under illumination the *I–V* characteristic shifts to lower voltage, with the light induced voltage shift being proportional to the injected optical power. It is also evident that the device's photo response is nonlinear, with the responsivity being a function of the bias voltage and the optical power. The measured responsivity was about 10 A W^{-1} at biasing close to the peak voltage and for injected light power of 0.1 mW. The responsivity dropped to 5 A W^{-1} when the optical power was increased to 1 mW. These results confirm that the responsivity is a highly nonlinear function of the optical power.

The device's response to 1550 nm wavelength light, shown in figure 5.9(b), was measured in the same conditions and power levels. The maximum measured responsivity was 3 A W^{-1} for 0.1 mW and 2 A W^{-1} for 1 mW optical powers, respectively. The lower values of responsivity obtained at 1550 nm are mainly associated with the lower absorption coefficient of the InGaAs at 1550 nm, 0.68 μm^{-1} at 1550 nm when compared with 1.16 μm^{-1} at 1310 nm.

Next, we present the characterisation of the InGaAlAs/InP RTD-PD epilayer incorporating double heterostructures, which can also be used to implement electro-absorption modulators for operation at around 1550 nm. The epilayer corresponds to a symmetric InGaAlAs/InP RTD-PD with two identical light sensitive layers adjacent to the DBQW, sandwiched by two InAlAs/InGaAs double heterostructures. The corresponding epilayer-stack is presented in table 5.2 and includes two 500 nm InGaAlAs (7% of Al) quaternary compound lattice matched to the InP each side of the DBQW, and two InAlAs layers lattice matched to the InP that act as cladding layers in the case of RTD-PD waveguide implementations. They are added to reduce the impact of the light absorption in the high-doped regions, which (as mentioned previously) deteriorate the response speed of the devices because the

Table 5.2. Symmetric double-heterostructure epilayer of InGaAlAs/InP RTD-PDs designed to operate in the wavelength range of 1200–1560 nm. The layer description assumes that the devices are forward biased.

Layer number	Thickness (nm)	Material	Doping (cm^{-3})	Description
1	100	$In_{0.53}Ga_{0.47}As$	2×10^{19} Si	Collector
2	300	$In_{0.52}Al_{0.48}As$	2×10^{18} Si	Collector
3	500	$In_{0.53}Ga_{0.40}Al_{0.07}As$	5×10^{16} Si	Absorption
4	2	$In_{0.53}Ga_{0.40}Al_{0.07}As$	Undoped	Spacer
5	1.7	AlAs	Undoped	Barrier
6	5.7	$In_{0.53}Ga_{0.47}As$	Undoped	Well
7	1.7	AlAs	Undoped	Barrier
8	2	$In_{0.53}Ga_{0.40}Al_{0.07}As$	Undoped	Spacer
9	500	$In_{0.53}Ga_{0.40}Al_{0.07}As$	5×10^{16} Si	Absorption
10	300	$In_{0.52}Al_{0.48}As$	2×10^{18} Si	Emitter
11	500	$In_{0.53}Ga_{0.47}As$	2×10^{19} Si	Emitter
12	100	$In_{0.53}Ga_{0.47}As$	Undoped	Spacer
13	20	InP	Undoped	Buffer
		InP substrate		

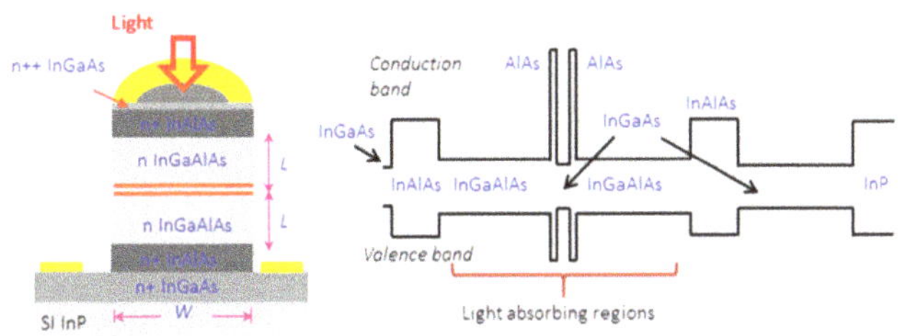

Figure 5.10. Physical layout of InGaAs/InP RTD-PD mesas with light-absorbing layers 500 nm thick both sides of the DBQW. A representation of the bottom of the conduction energy and top of the valence band energy profiles versus distance for the device at zero bias is also shown.

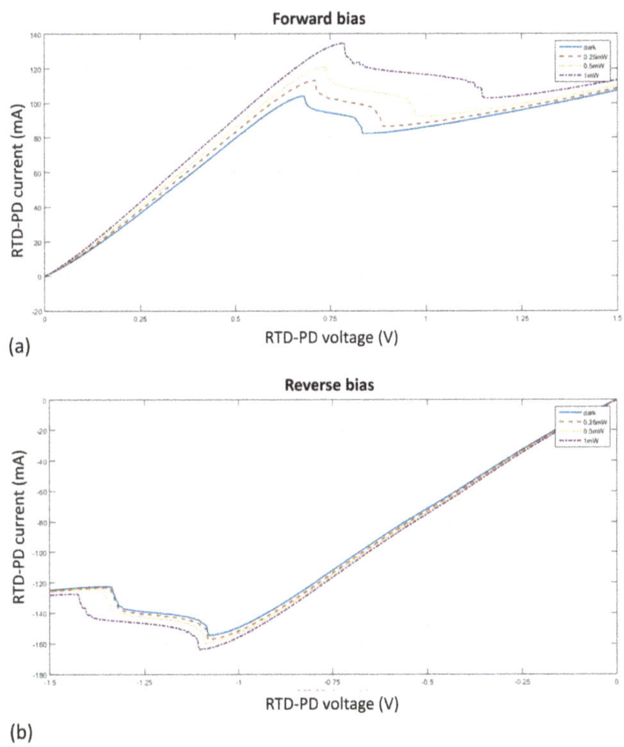

Figure 5.11. Measured current-voltage response of mesa InGaAlAS/InPRTD-PD illuminated at a wavelength of 1310 nm under (a) forward and (b) reverse bias, respectively.

photogenerated carriers in these regions will move through the diffusion transport process, which is a relatively slow transport mechanism for charge carriers.

Figure 5.10 shows a schematic of the physical layout of double-heterostructure InGaAs/InP RTD-PD of table 5.2. A representation of the conduction and valence band-edge profiles at zero bias is also shown.

Vertically illuminated RTD-PDs 25 μm × 25 μm in size with 15 μm × 15 μm optical windows were tested. Figure 5.11(a) presents the InGaAlAs/InP RTD-PD

I–V characteristic under forward bias at dark conditions (blue line), and when illuminated with 1310 nm light for optical powers: 0.25 mW (red line), 0.5 mW (yellow line) and 1 mW (purple line). The measured responsivity was 28.8 and 15.1 A W^{-1} on the first and second PDC, respectively, for 0.25 mW illumination; 24.4 and 13.5 A W − 1 on the first and second PDC, respectively, for 0.5 mW; 19.7 and 10.8 A W^{-1} on the first and second PDC, respectively, for 1 mW.

The light induced *I–V* shift to lower voltage was also quantified as the light *voltage sensitivity* measured as the peak voltage shift per unit of optical power, measured in V W^{-1}. The measured light voltage sensitivity was 204.8 and 408.4 V W^{-1} on the first and second PDC, respectively, for 0.25 mW; 144.4 and 318 V W^{-1} on the first and second PDC, respectively, for 0.5 mW; 102.2 and 262 V W^{-1} on the first and second PDC, respectively, for 1 mW.

The RTD-PD photo response under reverse bias is shown in figure 5.11(b): *I–V* characteristics under dark conditions (blue line), and under illumination for 0.25 mW (red line), 0.5 mW (yellow line) and 1 mW (purple line). The measured responsivity was 10.8 and 2.4 A W^{-1} on the first and second PDC, respectively, at 0.25 mW; 10 and 3.6 A W^{-1} on the first and second PDC, respectively, at 0.5 mW; and 8.5 and 4.2 A W^{-1} on the first and second PDC, respectively, for 1 mW illumination. The lower responsivity values can be attributed to the reduced effectiveness of the applied voltage to induce the bands bending to achieve the resonant condition: the peak voltage occurs at higher voltage (well above 1 V, against 0.7 V for forward bias) with the peak current 50% higher when compared with the forward bias value. The light induced *I–V* shift to higher voltage was quantified as follows: 204 and 180 V W^{-1} on the first and second PDC, respectively, at 0.25 mW, 138 and 216 V W^{-1} on the first and second PDC, respectively, for 0.5 mW, 84 V W^{-1} on the first PDC when illuminated with 1 mW.

As shown in figure 5.11, for symmetric double heterojunction structures, the illumination induces an overall *I–V* curve shift towards higher voltage that increases with optical power. This *I–V* characteristic global shift is mainly attributed to the stronger impact of the photogenerated carriers in the 500 nm emitter spacer layer, when compared with the effect of hole accumulation in the 500 nm collector layer.

The photodetection response of an InGaAs/InP RTD-PD to 1310 nm modulated light at 1 GHz is shown in figure 5.12 with the optical power as a parameter. The inset shows the scope trace of the photo-detected signal. The highest responsivity obtained for these devices was 3.3 A W^{-1}. If we consider a PIN-PD with an equivalent intrinsic layer 1 μm thick with a quantum efficiency equal to 1 (not a realistic value, but quantum efficiencies up to 0.7 easily are achievable), then the theoretical responsivity would be below 1 A W^{-1} within the 1100–1600 nm wavelength range. Here we are not considering the Fresnel losses that can be minimised and we are assuming a coupling efficiency equal to 1, i.e., all arriving light is coupled to the photodetector.

More recently, we have investigated two asymmetric InGaAlAs/InP RTD-PD epilayers in double-heterostructure configuration with light sensitive layers adjacent to the DBQW with different thicknesses. These epilayers were thought to quantify the impact of the locations and the size of the light-absorbing layers relatively to the

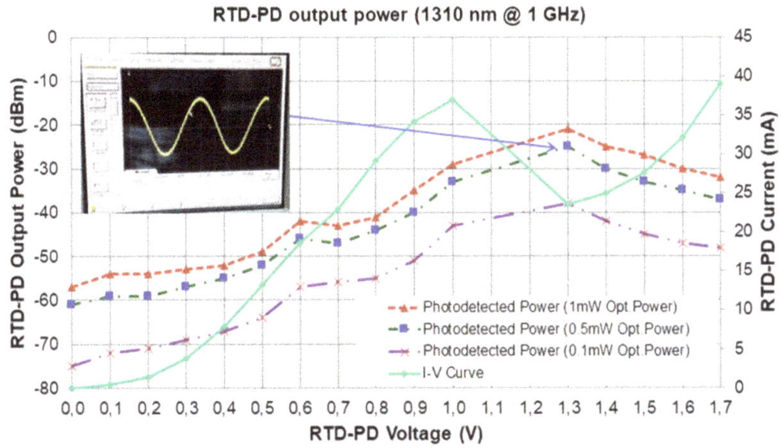

Figure 5.12. Response of InGaAs/InP RTD-PD to a 1310 nm light wave modulated at 1 GHz with the optical power as a parameter. The inset shows the scope photo-detected signal.

Table 5.3. Asymmetric double-heterostructure epilayer of InGaAlAs/InP RTD-PDs designed to operate in the wavelength range of 1200–1600 nm. The layer description assumes that the devices are forward biased.

Layer number	Thickness (nm)	Material	Doping (cm^{-3})	Description
1	100	$In_{0.53}Ga_{0.47}As$	2×10^{19} Si	Collector
2	100	$In_{0.52}Al_{0.48}As$	2×10^{18} Si	Collector
3	100/250	$In_{0.53}Ga_{0.47}As$	5×10^{16} Si	Absorption
4	2	$In_{0.53}Ga_{0.47}As$	Undoped	Spacer
5	1.7	AlAs	Undoped	Barrier
6	5.7	$In_{0.53}Ga_{0.47}As$	Undoped	Well
7	1.7	AlAs	Undoped	Barrier
8	2	$In_{0.53}Ga_{0.47}As$	Undoped	Spacer
9	20	$In_{0.53}Ga_{0.47}As$	5×10^{16} Si	Absorption
10	100	$In_{0.52}Al_{0.48}As$	2×10^{18} Si	Emitter
11	500	$In_{0.53}Ga_{0.47}As$	2×10^{19} Si	Emitter
12	100	$In_{0.52}Al_{0.48}As$	Undoped	Spacer
13	20	InP	Undoped	Buffer
		InP substrate		

DBQW. The corresponding epilayer stacks are presented in table 5.3. The epilayer stacks include a 20 nm InGaAs layer before the DBQW. They differ by the thickness of the InGaAs light sensitive layer that follows the DBQW: the thicknesses are 100 nm and 250 nm, respectively. The double heterostructures include (the same as the previous one) two InAlAs layers lattice matched to InP (here they were made 100 nm thick instead of 300 nm as in the symmetric RTD-PD).

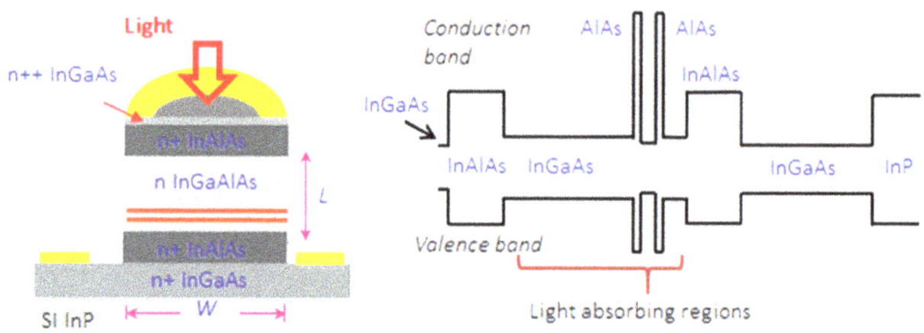

Figure 5.13. Physical layout of InGaAs/InP RTD-PD with asymmetric light-absorbing layers: 20 nm before and 100 nm/250 nm after the DBQW (following the growth direction), respectively. A representation of the conduction valence band-edges energy profiles at zero bias is also shown.

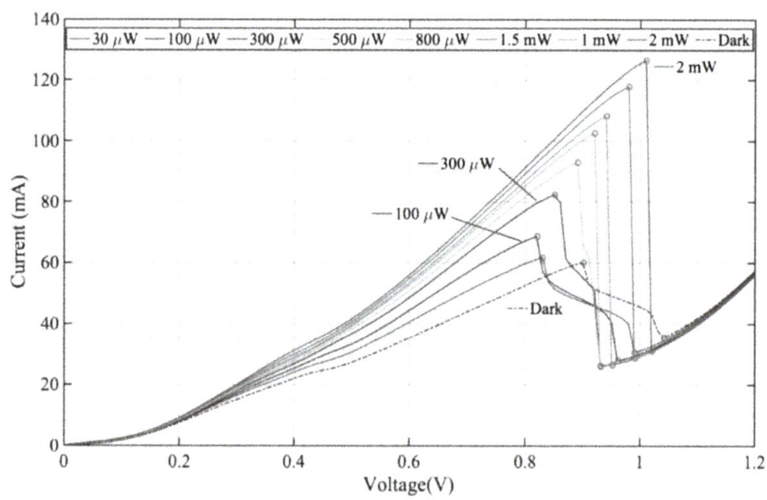

Figure 5.14. Photoresponse of InGaAs/InP RTD-PDs with 250 nm spacer: *I–V* for different optical powers (1550 nm). Reproduced from [16]. CC BY 4.0.

Figure 5.13 shows a schematic of the physical layout of the asymmetrical double-heterostructure InGaAs/InP RTD-PD of table 5.3. A representation of the conduction and valence band-edge profiles at zero bias is also shown. The devices fabricated from both epilayers only show N-shape *I–V* under reverse bias, meaning the 100/250 nm thick InGaAs layers now act as emitter spacer layers and the 20 nm InGaAs layer plays the role of collector spacer.

Figure 5.14 shows the impact of light illumination on the *I–V* of the 250 nm epilayer RTD-PD devices. The analysis of the *I–V*s shows that as the optical power increases, the *I–V* undergoes a shift initially to lower voltage until the optical power reaches a value around 0.1 mW. As the optical power rises above this threshold of around 0.1 mW, the *I–V* initiates a shift to higher voltage.

This behaviour can be explained as follows: for optical power below 0.1 mW, the effect of the photogenerated hole accumulation on the collector spacer (which for the epilayers in consideration under reverse bias corresponds to the 20 nm thick InGaAs layer) overcomes the effect of the photogenerated electron accumulation in the emitter. The overall effect then leads to the reduction of the applied voltage required to achieve the peak current, which results in an *I–V* shift to lower voltage.

When the optical power is higher than 0.1 mW, the increase of the conductivity due to the electron–hole pair photogeneration and the effect of photogenerated electron accumulation in the emitter adjacent to the DBQW overcomes the effect of the photogenerated hole accumulation in the collector. A larger voltage is now needed to achieve the required emitter band bending to reach the condition of maximum electron transmission (peak current). This leads to a current increase and to an overall *I–V* shift to higher voltage.

The measured responsivity of these devices was about 240 A W^{-1} at a bias of around 0.76 V for an optical power around 0.03 mW. For optical power levels on the mW range, the responsivity was about 20 A W^{-1}. Moreover, the responsivity rises non-linearly with the reduction of the optical power. There have been reports of responsivity values as high as 10 MA W^{-1} for illumination power on the order of few tens of picowatt [15].

RTD-PD devices with 100 nm spacers with identical size to the 250 nm spacer RTD-PDs show a much less pronounced effect, with an optical power turning point value less than 0.1 mW. In this epilayer, the photogenerated electron accumulation in the emitter spacer has a stronger effect on the electric field profile than the photogenerated hole accumulation in the collector spacer adjacent to the DBQW. The conductivity increase overcomes the impact of photogenerated hole accumulation in the collector at much lower power. For a detailed discussion, see [15, 16]. The responsivity of these devices reaches a value of 25 A W^{-1} at around 0.76 V for an optical power of 0.03 mW. The responsivity for optical power levels on the mW ranges is of the order of 7 A W^{-1}.

The RTD-PD photo response has a nonlinear behaviour that depends on the DC bias voltage and on the injected optical power, so that $\mathcal{R}_{\mathrm{RTD-PD}}$ becomes dependent on the applied bias voltage and optical power. Due to the built-in electrical field dependence on the charge distribution, the devices' responsivity (the ratio of the photocurrent and the absorbed optical power; photocurrent is the difference of the current levels under illumination and dark) is a nonlinear function of the bias voltage and the light irradiance. The overall photocurrent is a nonlinear function of the applied voltage because the device conductivity enhancement induced by the illumination is a nonlinear function of the applied bias.

The results from other groups show that at a given bias voltage under low power illumination, the photocurrent can be represented as a linear function of the incident optical power, which makes it possible to operate RTD-PDs as photon counters [14]. For a given current value, the amount of voltage shift depends on the applied voltage and on the optical irradiance being absorbed by the device. With optimised structures and maximising coupling of light into the absorption region, responsivities as high as $\mathcal{R} = 3.1 \times 10^4$ A W^{-1} have been measured with a sensitivity of several pA per photon [8].

When operated in the first PDC (passive mode) where the bias voltage is set to maximise the value of $\mathcal{R}_{RTD-PD}$, effective responsivity can be well above the APDs gain. The preliminary results discussed here seem to indicate that RTD-PDs can be designed to operate as photodetectors with superior responsivity values than PIN-PDs. The results also appear to confirm that RTD-PD structures can be designed to have responsivity values well above APDs when operating at very low levels of irradiances (below microwatt) [8, 14–16].

5.4.2 RTD-PD response time

We have performed the characterisation of the response time of RTD-PDs circuits. Preliminary results of the response time are summarised in figure 5.15. The devices were excited with 1 ns long optical pulses. The RTD-PDs were biased well below the peak voltage (in range 100–200 mV). The average fall times are in the range of tens of ns. It was observed that the RTD-PDs response times and decaying times do not change much with the applied voltage for bias below the peak voltage.

These preliminary results seem to indicate that the response times are being degraded by the diffusion of the photogenerated carriers in 500 nm InGaAs lower contact layer because the estimated drift, transit, and RC times for these devices are in the sub-nanosecond range. To test this hypothesis, we have measured the response time of two additional epilayer stacks with InGaAs light sensitive layers and the DBQW embedded between high-doped InP layers. The measured response times were in the nanosecond range, much faster than the times shown in figure 5.15 [15].

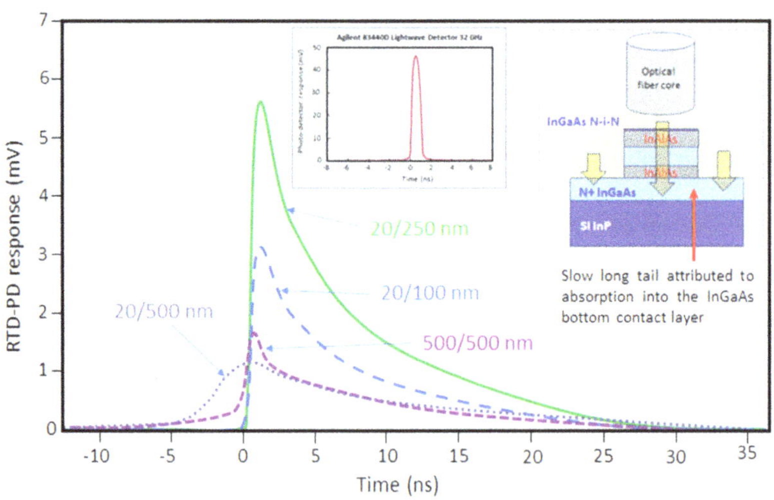

Figure 5.15. InGaAs/InP RTD-PDs response time to 1 ns long optical pulses (1550 nm). The inset shows the response time of a 32 GHz InGaAs PIN photodetector. A schematic picture of the light path is also shown, pointing out that a significant amount of light is absorbed in the lower 500 nm InGaAs layer used to contact formation.

The main difference from these epilayers to those being considered here is that they do not have the 500 nm high-doped InGaAs contact layer; instead, the bottom contacts are made on InP. This 500 nm InGaAs layer absorbs a significant amount of light, generating a large number of e–h pairs, because the high doping these photogenerated e–h pairs experience a very weak electric field. Consequently, they contribute to the current through diffusion, which is a slow transport process when compared to the drift under high-electric field (electric field >1 kV cm^{-1}).

The results obtained so far seem to imply that well-designed RTD-PD structures may have the potential to achieve both very high responsivity and sub-nanosecond response times. These characteristics can make RTD-PD performance overcome PIN-PDs and APDs in the main metrics required for low cost high-speed optical communication systems, including radio-over-fibre communication systems, as well as receivers and spike generators in photonic SNNs, as photon counters and as single-photon detectors for quantum-based technologies.

The resonant tunnelling process has also been employed using quantum dots for single-photon detectors, albeit at cryogenic temperatures [4]. The combination of the responsivity and the noise level determines the sensitivity, which is the smallest amount of energy that a photodetector can transmit with a given signal-to-noise ratio [4]. For RTD-PD, the excess noise factor, F, has not been extensively studied. There has been some work on the shot noise of the current flowing through an RTD and there is evidence that, as a consequence of the quantum tunnelling process, the current shot noise through a RTD is suppressed [20].

5.5 RTD-PD excitability and optically induced spike generation

Both RTD-PDs and APDs are types of excitability detectors, i.e., detectors that can deliver a large pulse output for a small trigger input but then have a dead time while the external circuit recharges the device, so that they can then deliver another output pulse when they receive another trigger input [17–19, 23].

The concept of excitability originates in biological systems and the operation of neurons, and provides useful insights for computing systems that emulate neural networks [35]. Neurons are excitable units that produce spikes or bursts of electrical signals: while resting in their stable states, they produce pulses of fixed shape and amplitude whenever they are excited beyond a certain threshold. If the perturbation is below the threshold, then they either do not react or produce a much smaller response. These distinguish responses: the excitable pulses and negligible pulse are known as all-or-nothing response [17–19].

RTD-PDs can operate as excitability photodetectors and be used to build optoelectronic neuromorphic receiver circuits that are capable of producing optically induced electric spiking events with nanosecond repetition rate. Figure 5.16 shows a schematic of an excitable RTD-PD circuit. Figure 5.16(a) shows a diagram of the circuit used in characterisation of RTD-PD excitability properties (left) and the equivalent circuit of an RTD-PD oscillator circuit (right).

For operation as a single-pulse generator, i.e., as a spike generator, the circuit must operate as a monostable circuit. A monostable circuit is a circuit that has one

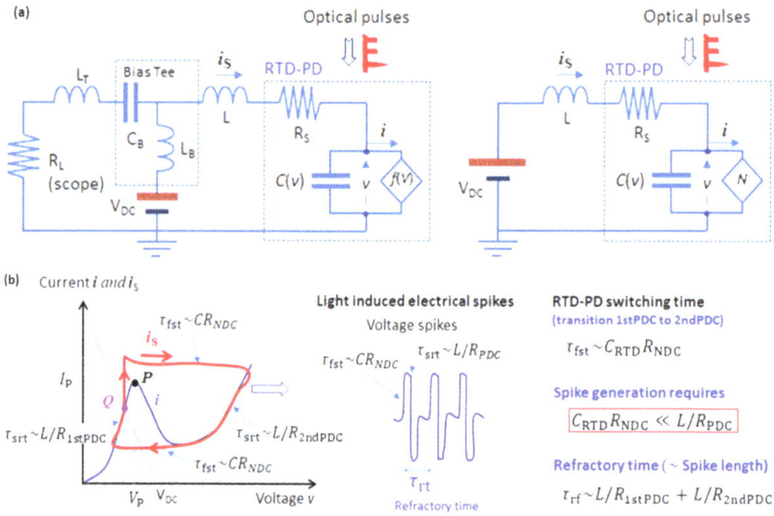

Figure 5.16. (a) Diagram of the circuit used in characterisation of RTD-PD excitability properties, and equivalent circuit of an RTD-PD oscillator circuit. (b) Schematic of the RTD-PD *I–V* curve and the circuit load line, showing the quiescent-point trajectory when the optical pulse's intensity exceeds a given threshold. The conditions needed to observe spike generation, with the circuit relevant time constants are also shown.

stable state and one quasi-stable state. To operate as a monostable circuit, the RTD-PD and the circuit must be designed in such a way that the load line intersects the RTD-PD *I–V* just in a single point and that point must be in the PDC portion of the *I–V* not far from the NDC region. Figure 5.16(b) presents the schematic of the generic path of the circuit quiescent point when the intensity of an optical pulse overcomes the firing threshold and induces the transition from the stable state to the quasi-stable state, followed by a quick return to the stable state.

The operation of an RTD-PD circuit as an excitable system is as follows. When an optical pulse impinges the light sensitive region of the device, the pulse absorption leads to the generation of electron–hole pairs whose dynamics pulls the voltage v and the current i_s from the stable operation point Q by the amounts Δv and Δi_s. If the pulse intensity is below the threshold, then the voltage and current changes, Δv and Δi_s, are small quantities that decay towards the stable point values not being able to shift the load line to the NDC region. The operation point returns to the stable point Q with a time constant given by ~ $L/(R_S + R_{1PDC})$, where R_{1PDC} is the positive differential resistance of the first PDC portion of the *I–V* curve.

Whenever the pulse intensity is above a given threshold voltage and current changes, Δv and Δi_s, are such that the load line is moved into the NDC. Since this operating point is instable it induces the RTD-PD circuit operating point to switch to a quasi-stable state bias point in the second PDC. The quiescent-point switching time from the first PDC to the second PDC is quite fast, given by ~$5CR_{NDC}$ (similarly to the switching time of an RTD), where C is the RTD-PD capacitance and R_{NDC} is the negative differential resistance of the *I–V* curve. This RTD-PD

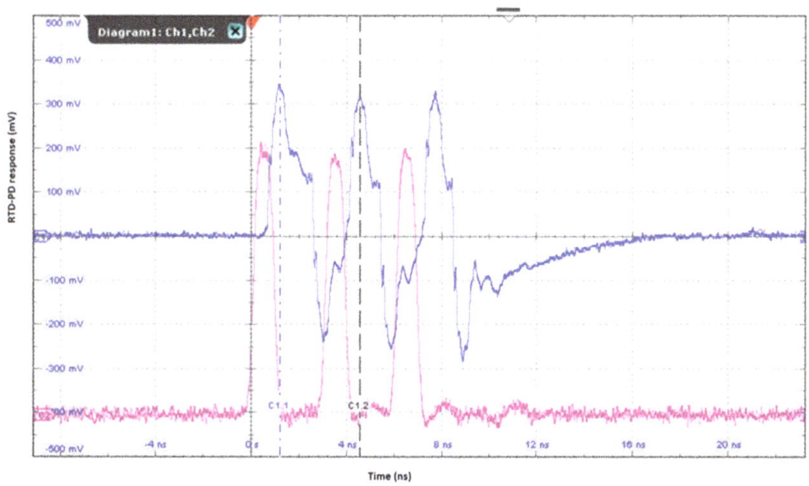

Figure 5.17. RTD-PD spike (blue) response to three consecutive 1 ns optical pulses (red-purple).

switching between PDCs is accomplished with a fast voltage rise while the current does not change significantly, as sketched in figure 5.16(b).

Once in the quasi-stable state in the second PDC, the voltage and the current starts to drop at a much slower rate than the RTD switching time between PDCs. The time constant of the current and voltage decrease in the second PDC is given by $\sim L/(R_S + R_{2PDC})$, where L is the circuit equivalent inductance and R_{2PDC} is the positive differential resistance in the second PDC of the I–V curve. The voltage and current drops until the load line crosses the valley point towards the NDC region. The operation point then returns very quickly to the first PDC portion of the I–V (again in a time $\sim 5CR_{NDC}$), with a significant voltage drop and a negligible change in current, see figure 5.16(b). From there, the operation point returns to its initial position (stable point), which increases the voltage and current with a time constant given by $\sim L/(R_S + R_{1PDC})$. This behaviour is shown in figure 5.17 for an RTD-PD excitable circuit using the InGaAs/InAlAs RTD-PDs with the 250 nm intentionally added InGaAs spacer layer. This device response was optically driven by a sequence of three consecutive 1 ns long optical pulses. In figure 5.17, the blue trace is the RTD-PD spike response and the red-purple trace is the photodetection response of a 32 GHz bandwidth PIN-PD (Agilent) to the three consecutive 1 ns long optical pulses. These devices show similar behaviour when using sequences of two consecutive pulses, producing two spikes.

When the consecutive optical pulses were less than $\approx$2.5 ns apart, the circuit generated a single spike per each sequence of consecutive pulses. This minimum time separation corresponds to the refractory or restoring time of the RTD-PD excitable circuit, i.e., the minimum time between consecutive perturbations the circuit requires to be able to generate two consecutive light induced independent spikes. For this RTD-PD circuit, the refractory time is given by the sum of the two slow time constants associated to the voltage and current changes across the PDC regions, as indicated in figure 5.16. For a more detailed description of this behaviour, see [17, 18, 35]. The RTD-PD excitable circuit may have potential to function as photon counters and single photodetectors [14].

5.6 Optical control of high-frequency RTD oscillators

High-speed photonic-microwave interfaces are a key technology for the coming generations of higher capacity radio-over-fibre based wireless communication systems, 5G and above. The requirement for tens of GBps data rates per user needs a concomitant increase in carrier frequency, resulting in systems using carrier frequencies of hundreds of gigahertz (GHz). Although synthesised multiplier-based systems can be used to achieve these high frequencies, they require many stages of multiplication, which results in high power consumption and large size. An alternative approach is to use advanced solid-state oscillators with a fundamental frequency in the terahertz range as the carrier frequency.

As mentioned previously, RTDs are currently the basis of the highest frequency solid-state oscillator, and therefore they have the potential to satisfy the requirements of the next generations of radio-over-fibre based wireless communication systems. Indeed, RTD-PDs are thought to be potential candidates for the low cost integration of radio and fibre networks. In this section, we discuss the capabilties of RTD-PD oscillators to operate as optical-to-microwave transceivers in wireless data links [13, 36]. The concept of the RTD-PD optical-to-wireless data link is illustrated in figure 5.18. The RTD-PD is configured as an electronic oscillator. The phase and/or the amplitude of the RTD oscillator output is controlled by an optical signal.

As described in [36], an RTD oscillator circuit comprising an DBQW AlAs/InGaAs/AlAs RTD-PD with a InGaAlAs optical absorption layer was used as the RTD-PD for the optical wireless link. The bandgap of the InGaAlAs alloy was adjusted so that light at optical communications wavelengths (in the range 1200 nm to 1600 nm) was strongly absorbed in the light absorption layers. The RTD-PD circuit oscillation frequency was set to 34 GHz by the intrinsic RTD-PD capacitance and the inductance of the coplanar waveguide (CPW) transmission line. When DC biased to the appropriate voltage, in this case in the range ~0.7–1.05 V, the RTD-PD oscillated at 34 GHz and could be amplitude modulated, phase modulated and

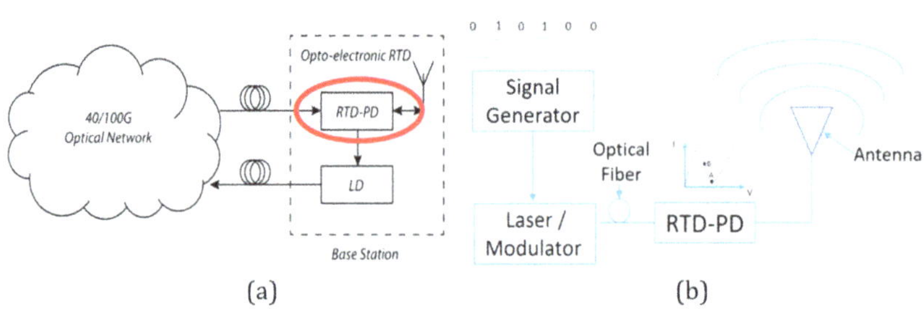

Figure 5.18. (a) Schematic representations of RTD-PD application on radio-over-fibre wireless communication systems. The RTD-PD is configured as a photodetector with built-in gain and as an electronic oscillator. The phase and/or the amplitude of the RTD-PD oscillator output can be controlled by both an electrical and an optical signal. (b) Schematic of optically controlled RTD-PD configured as an optical-to-wireless data link.

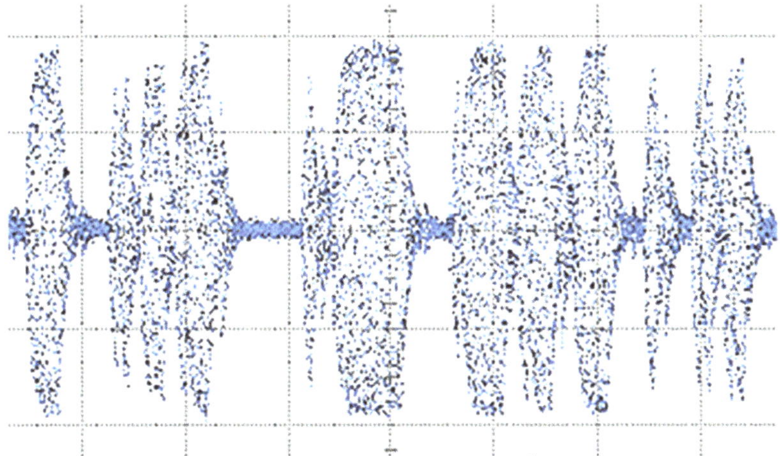

Figure 5.19. The output signal in the time domain from the RTD-PD showing data patterns at 80 Mbit s^{-1} amplitude modulating a RF carrier wave at 34 GHz.

frequency modulated by a combination of the applied optical signal and the exact setting of the DC bias voltage.

Figure 5.19 shows the RTD-PD output signal in the time domain of the amplitude modulated at a data rate of 80 Mbit s^{-1}. Quadrature phase shift keying with a similar set-up with data rates up to 200 Mbit s^{-1} was also demonstrated. Subsequent work [13] has operated with higher carrier frequencies of up to 79 GHz and data rates of 50 Mbit s^{-1}.

As described in [13], injected modulated RF signals and intensity modulated optical signals were able to shift the bias operation point of a RTD-PD circuit with fundamental frequency of oscillation at 79 GHz but biased at the end of the first PDC region, just before the NDC region, from the steady state (PDC region) to the oscillatory state (NDC region) to generate a mm-wave signal that transfers the input data signal onto an RF carrier.

In principle, this technique can be scaled to higher data rates and higher carrier frequencies, and when used with integrated antennae can result in very low cost, high-performance flexible electronic, optical and radio interfaces.

5.7 Design guidelines for high-bandwidth RTD-PDs

As discussed, the position of the light-absorbing layer(s) relative to the DBQW also determines the effect of the light-generated electron–holes pairs on the *I–V* characteristic and on the optoelectronic response including the photocurrent–voltage and photocurrent–irradiance characteristics, and of course the response time. To obtain a high-speed response, the light sensitive layers must be on the collector spacer where the electric field distribution reaches it maximum values. To maximise the responsivity, it is more favourable to locate the light sensitive layers on the emitter's side.

The RTD-PD photocurrent generated by high-frequency modulated optical signals is smaller than in the stationary state because of bandwidth limiting effects, such as the carriers' diffusion time, RC-time and transit-time constraints. The lifetimes of photo-generated and accumulated charge carriers set the upper bounds for the device sensitivity and operation frequency [26]. To achieve high-performance RTD-PDs, such as RTD-PDs with simultaneous high responsivity and high-speed response, some improvements must be considered on their current epilayer structures.

For vertically illuminated RTD-PD structures, the light propagates perpendicularly to the plane defined by the absorbing layer (and the heterojunction's interfaces). Therefore, the bandwidth-quantum efficiency product depends on the operating wavelength (through the absorption coefficient). There is a trade-off between the cutoff frequency and the quantum efficiency, resulting in a limitation on the bandwidth-efficiency product—this is generally true for any VIPD. Another limitation arises from the trade-off between bandwidth and saturation power. A high bandwidth (high cutoff frequency) requires, in general, a small absorption volume, which results in a high carrier concentration at a given power level for the optical signal, which can lead to saturation. As a numerical reference and assuming single-pass absorption, it is estimated that for an absorbing coefficient of the order of 1 μm^{-1}, the ideal bandwidth-efficiency product of a PIN-PD in the third and second optical communication windows will be below 20 and 35 GHz, respectively [26].

To eliminate or substantially reduce the carrier-diffusion time, the RTD-PD structure should avoid the presence of any light sensitive layer beyond the active region of the devices, i.e., the light sensitive layers must be confined to the regions adjacent to the DBQW and be sandwiched between high bandgap double heterostructures. They must be based on a double heterostructure with the contacts made on high bandgap materials, such as InAlAs or InP, and avoid the presence of light sensitive layers outside the photodetection active region to eliminate or reduce the photogenerated carriers' diffusion in regions where the electric field is negligible. A RTD-PD implemented using InGaAs/InP double heterostructures is schematically represented in figure 5.20.

In RTD-PD waveguide configuration (WGPD), the optical signal is coupled and guided by the light sensitive layers in the direction that is parallel to the

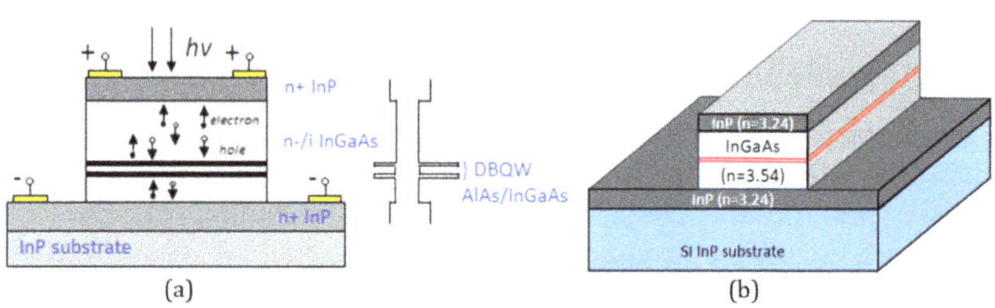

Figure 5.20. Schematic representation of the physical layout of double-heterostructure InGaAs/InP RTD-PDs: (a) an RTD-VIPD that minimises the diffusion of the photogenerated carriers, leading to superior response times; and (b) an RTD-WGPD that can maximise both responsivity (gain) and bandwidth-efficiency product.

heterojunction's interfaces. This direction is perpendicular to the flow of the photogenerated carriers. In general, waveguide photodetectors can have both a large bandwidth-efficiency product and a high saturation power because the waveguide geometry decouples the absorption length of the optical signal, i.e., the distance along the length of the active region, from the photogenerated carriers' flow path, which encompasses the full thickness of the light sensitive regions. Because the absorption length is no longer tied to the thickness of the light sensitive region, RTD-PD can have both a thinner active region to achieve short transit times and a long absorption length for attain high quantum efficiency and high saturation power.

Since an RTD-WGPD is formed by integrating the light sensitive region within an optical waveguide structure, there are two basic schemes for injection of light: the butt-coupling configuration and the evanescent-coupling configuration. In the butt-coupling scheme, also called the end-coupling or end-firing, the light sensitive region sits within or coincides with the waveguide core and a feeding waveguide, e.g., an optical fibre, is directly aligned with the RTD-PD. The RTD-PD waveguide can be made wide to support multiple transverse modes to improve the coupling efficiency. Since absorption takes place along the length of the optical waveguide, the quantum efficiency can be nearly 100% even for an ultrathin absorption layer.

In the evanescent-coupling scheme, the photosensitive region of the RTD-PD can be located outside the waveguide core (made of a non-absorbing material at the operating optical wavelength), typically on top of the waveguide, and the light is transferred to the detection active region through the resonant optical tunnelling effect (ROTE) [21, 26].

RTD-WGPD, akin the RTD electro-absorption modulator arrangement, can maximise the external quantum efficiency without degrading the photodetection speed response because the responsivity can be made to not depend on the thickness of the absorbing layer, and the photogenerated carrier-transit times can be independently tailored without sacrificing their quantum efficiency and saturation power.

The transit-time-limited bandwidth of RTD-WGPDs can be made larger than their RC-time-limited bandwidth by using sufficiently thin active regions. As a result, the primary bandwidth limitation will not be the carrier-transit time but the RC-time. Furthermore, RTD-WGPDs with thin active regions have a few additional advantages. A thin active region allows the device to operate at a lower bias voltage, resulting in a smaller dark current and reduced noise contributed by the dark current. The waveguide geometry is also compatible with other waveguide-based photonic devices and components, making it easy to incorporate RTD-based photodetectors into integrated photonic circuits with reduced input and output coupling losses.

5.8 Conclusions

Because RTDs can have large bandwidth NDC, they can easily be configured as a high-frequency electronic oscillator [1]. With the addition of light absorption layers, the RTD-PD can be used as a high-frequency electronic oscillator controlled by optical signals. This can form the basis of future low cost high-density optical-to-wireless data links. Preliminary results show the operation with carrier frequencies of up to 79 GHz and data rates of up to 200 Mbit s^{-1}. This approach clearly has the potential to go to high carrier frequencies and with much superior data rates because the RTD circuits have been shown to be the key active component in electronic oscillators operating at up to 450 GHz with powers of 10 mW [2] and at 1.92 THz with powers of 0.4 μW [1].

This chapter reviewed recent findings on the behaviour of RTD-PD structures in terms of responsivity and response time, and the impact of the light sensitive spacer layer's characteristics (position and thickness) on the performance of the device (responsivity and response time). RTD-PDs have highly nonlinear photocurrent–voltage (Iph–V) and photocurrent–light power (Iph–L) responses, which make them suitable for the detection of very weak optical signals, followed by built-in amplification. Their highly nonlinear responses make them suitable for digital communication systems but some of the metrics from linear photodetectors must be used carefully when describing RTD-PDs (and indeed APDs).

The results seem to indicate that high responsivity and fast response time may be achieved if double-heterostructure RTD-PDs are considered where all the light sensitive layers and the DBQW are sandwiched between the high bandgap regions of double heterostructures.

Additionally, RTD-PDs like APDs belong to a class of detectors that under certain operating conditions can be described using concepts from nonlinear dynamics, such as excitability. An investigation of RTD-PD circuits as excitable systems indicates that they can operate as light spike detectors and as light induced spike generator nodes in photonic SNNs. The nonlinear dynamics of RTD photonic circuits are comprehensively covered in the following chapters.

In conclusion, because of their built-in wide-bandwidth nonlinear electrical gain, RTD-PDs can amplify very weak photogenerated electric signals, providing an effective optical responsivity that is much superior to photoconductive detectors such as PIN-PD and APDs. Due to their simplicity and low voltage operation, they are being considered as potential high-speed optical-to-electrical transducers for future low cost light-wave and radio-over-fibre communication systems, as light spike detectors and as light induced spike generator nodes in photonic SNNs, and as photon counters and single-photon detectors for quantum-based technologies.

References

[1] Maekawa T, Kanaya H, Suzuki S and Asada M 2016 Oscillation up to 1.92 THz in resonant tunneling diode by reduced conduction loss *Appl. Phys. Express.* **9** 024101

[2] Suzuki S, Shiraishi M, Shibayama H and Asada M 2013 High-power operation of terahertz oscillators with resonant tunneling diodes using impedance-matched antennas and array configuration *IEEE J. Sel. Top. Quantum Electron.* **19** 8500108

[3] Koyama Y *et al* 2022 A high-power terahertz source over 10 mW at 0.45 THz using an active antenna array with integrated patch antennas and resonant-tunneling diodes *IEEE Trans. Terahertz Sci. Technol.* **12** 510–9

[4] Li H W, Kardynal B E, Ellis D J P, Shields A J, Farrer I and Ritchie D A 2008 Quantum dot resonant tunneling diode single photon detector with aluminum oxide aperture defined tunneling area *Appl. Phys. Lett.* **93** 153503
[5] Romeira B, Pessoa L M, Salgado H M, Ironside C N and Figueiredo J M L 2013 Photodetectors integrated with resonant tunneling diodes *Sensors* **13** 9464–82
[6] Dong Y *et al* 2015 Resonant tunnelling diode photodetector operating at near-infrared wavelengths with high responsivity *Electron. Lett.* **51** 1355–6
[7] Dong Y *et al* 2014 Resonant tunneling diode with a multiplication region for light detection *Opt. Commun.* **331** 94–7
[8] Pfenning A, Hartmann F, Langer F, Hofling S, Kamp M and Worschech L 2014 Cavity-enhanced resonant tunneling photodetector at telecommunication wavelengths *Appl. Phys. Lett.* **104** 101109
[9] Pfenning A *et al* 2015 Photocurrent–voltage relation of resonant tunneling diode photodetectors *Appl. Phys. Lett.* **107** 081104
[10] Rodrigues G C *et al* 2017 Resonant tunneling diode photodetectors for optical communications *Proc. SPIE* **10453**
[11] Nie B Y *et al* 2019 InAs/GaSb superlattice resonant tunneling diode photodetector with InAs/AlSb double barrier structure *Appl. Phys. Lett.* **114** 053509
[12] Pfenning A, Hartmann F, Weih R, Emmerling M, Worschech L and Höfling S 2018 p-Type doped AlAsSb/GaSb resonant tunneling diode photodetector for the mid-infrared spectral region *Adv. Opt. Mater.* **0** 1800972
[13] Zhang W *et al* 2019 Optical direct intensity modulation of a 79 GHz resonant tunneling diode-photodetector oscillator *Opt. Express* **27** 16791
[14] Pfenning A, Kruger S, Jabeen F, Worschech L, Hartmann F and Hofling S 2022 Single-photon counting with semiconductor resonant tunneling devices *Nanomaterials* **12** 2358
[15] Alomari S A Y 2022 Resonant tunnelling diode optoelectronic receivers and transmitters *PhD Thesis* (Ciências, Universidade de Lisboa)
[16] Alomari S, Al-Taai Q, Elksne M, Al-Khalidi A, Wasige E and Figueiredo J 2023 Speed limitations of resonant tunneling diode-based photodetectors *Opt. Express* **31** 18300–17
[17] Ortega-Piwonka I, Piro O, Figueiredo J, Romeira B and Javaloyes J 2021 Bursting and excitability in neuromorphic resonant tunneling diodes *Phys. Rev. Appl.* **15** 034017
[18] Romeira B *et al* 2023 Brain-inspired nanophotonic spike computing: challenges and prospects *Neuromorph. Comput. Eng.* **3** 033001
[19] Hejda M *et al* 2023 Artificial optoelectronic spiking neuron based on a resonant tunnelling diode coupled to a vertical cavity surface emitting laser *Nanophotonics* **12** 857–67
[20] Aleshkin V Y *et al* 2004 Giant suppression of shot noise as signature of coherent transport in double barrier resonant diodes *Semicond. Sci. Technol.* **19** S161–3
[21] Chuang S L 2010 *Physics of Optoelectronic Devices* (New York: Wiley)
[22] Campbell J C 2007 Recent advances in telecommunications avalanche photodiodes *J. Lightwave Technol.* **25** 109–21
[23] Campbell J C 2016 Recent advances in avalanche photodiodes *J. Lightwave Technol.* **34** 278–85
[24] Saleh B E A and Teich M C 2019 *Fundamentals of Photonics* (New York: Wiley)
[25] Agrawal G P 2010 *Fiber-Optic Communication Systems* (New York: Wiley)
[26] Weber H-G and Nakazawa M 2007 *Ultrahigh-Speed Optical Transmission Technology* (Berlin: Springer)

[27] Slight T J and Ironside C N 2007 Investigation into the integration of a resonant tunnelling diode and an optical communications laser: model and experiment *IEEE J. Quantum Electron.* **43** 580–7

[28] Romeira B, Figueiredo J, Ironside C N and Javaloyes J 2013 Dynamics of lienard optoelectronic oscillators *Sel. Topics Nonlinear Dynam. Theor. Electr. Eng.* **459** 117–38

[29] Romeira B *et al* 2009 Nonlinear dynamics of resonant tunneling optoelectronic circuits for wireless/optical interfaces *IEEE J. Quantum Electron.* **45** 1436–45

[30] Rockwell A K *et al* 2018 Toward deterministic construction of low noise avalanche photodetector materials *Appl. Phys. Lett.* **113** 102106

[31] Hayat M M *et al* 2002 Gain-bandwidth characteristics of thin avalanche photodiodes *IEEE Trans. Electron Devices* **49** 770–81

[32] Nakajima F, Nada M and Yoshimatsu T 2016 High-speed avalanche photodiode and high-sensitivity receiver optical subassembly for 100-Gb/s ethernet *J. Lightwave Technol.* **34** 243–8

[33] Semiconductor Database. Ioffe Institute. (http://www.ioffe.ru/SVA/NSM/Semicond/) (accessed 1 August 2023)

[34] Personick S D 1973 Receiver design for digital fiber optic communications systems 1 *Bell Syst. Tech. J.* **52** 843–74

[35] Yu T, Sejnowski T J and Cauwenberghs G 2011 Biophysical neural spiking, bursting, and excitability dynamics in reconfigurable analog VLSI *IEEE Trans. Biomed. Circuits Syst.* **5** 420–9

[36] Watson S *et al* 2019 Resonant tunneling diode photodetectors for optical communications *Microwave Opt. Technol. Lett.* **61** 1121–5

Charlie Ironside, Bruno Romeira and José Figueiredo

Chapter 6

Resonant tunnelling diode—laser diode (RTD—LD and LEDs)

Chapter 6 will examine the combination of resonant tunnelling diodes (RTDs) with lasers and light-emitting diodes (LEDs). In the resonant tunnelling diode-laser diode (RTD-LD) device the RTD provides wide bandwidth electronic gain, which enables a semiconductor laser to be rapidly modulated by a relatively small signal. For the resonant tunnelling diode-light-emitting diode (RTD-LED), the RTD provides a mechanism to creating electron–hole pairs which is markedly different from the conventional LED. The RTD-LED does not require a *pn* junction. Electrons are promoted to the conduction band via an impact ionisation mechanism, thus creating electron–hole pairs that can recombine via spontaneous emission of light.

6.1 Introduction

For III–V semiconductor devices, there is considerable convergence between electronics and optoelectronics regarding epitaxial layer material growth, device fabrication techniques and device layout. This has prompted investigations into integrating semiconductor light-emitting devices with purely electronic devices to produce optoelectronic integrated circuits (OEICs) with expectations of emulating the gains in functionality, speed and compactness that have been so significant in developing integrated silicon electronics. There has been work on RTD light-emitting diode integration [1], RTD laser diode integration [2–4], Gunn diode and laser integration [5], and transistor laser integration [6].

As noted in chapter 5, the resonant tunnelling diode photodetector (RTD-PD) has large enough electrical gain (or responsivity) to directly drive another device, such as a semiconductor laser. This has been an essential feature for the development of the RTD-based optoelectronic oscillator (OEO), which will be the subject of chapter 8. In this chapter, we discuss the development of the RTD-LD, with

doi:10.1088/978-0-7503-5714-2ch6

particular emphasis on it as a component of the OEO. Additionally, this chapter will review the recent developments of integrated RTD-LED devices.

6.2 Hybrid RTD laser circuit

The hybrid RTD-LD consists of a separate laser diode and RTD chips that are electrically coupled via a short (~2 mm) transmission line (gold (Au) wire bond), as illustrated in figure 6.1.

Compared to a fully integrated device, an advantage of the hybrid approach is that it is possible to separately characterise the current–voltage (*I–V*) properties of the laser and RTD chips and then combine them to investigate how the *I–V* curve changes, as illustrated in figure 6.2.

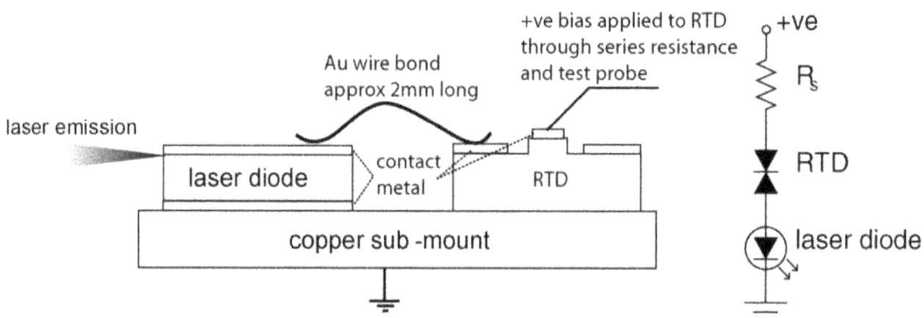

Figure 6.1. Illustration of the RTD-LD hybrid integration module. The RTD and laser dies are attached to the copper sub-mount with silver epoxy, a wire bond is made between them using a wedge bonder. On the right-hand side is a schematic lumped element circuit representation of the device. © [2007] IEEE. Reprinted, with permission, from [3].

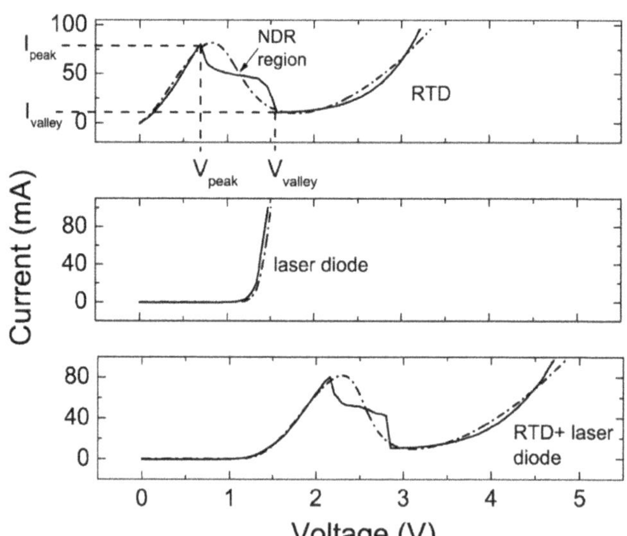

Figure 6.2. The *I–V* characteristics of the RTD-LD: the top panel is the RTD chip, the middle is the laser diode chip and the bottom panel is the hybrid RTD plus laser circuit. The solid lines are the measured data and the dashed lines are fitted from a PSPICE model. © [2007] IEEE. Reprinted, with permission, from [3].

The RTD I–V curves illustrated in figure 6.2 have a region indicated as the negative differential resistance (NDR; related to the NDC) region. In this flat plateau region, the hybrid circuit is oscillating and the oscillation frequency is determined by the value of the external parasitic components that are introduced by the bias circuit.

6.2.1 Model of the RTD-LD hybrid circuit

Initially, the RTD-LD was modelled using an established circuit software simulation tool called PSICE. However, more insight into the device's operation was obtained with the Liénard oscillator, which is a concept from the nonlinear dynamics community [7]. The details of this Liénard oscillator model are covered in chapters 7 and 8 but in broad outline the Liénard oscillator is a generalisation of the more well-known Van der Pol oscillator model, which was first used in the 1920s to describe devices and systems that exhibited negative differential resistance. The Van der Pol oscillator uses a third-order polynomial to represent the I–V curve of an electronic oscillator—it is the lowest order polynomial that has an NDC feature and so the Van der Pol oscillator is the simplest model of an NDC oscillator [8]. However, the Liénard oscillator model uses higher order polynomials that can more accurately represent the NDC curve of any given device or system. For our application, i.e., the RTD-LD hybrid circuit, the Liénard oscillator resulted in a more accurate model of the system and allowed the oscillation output of the RTD-LD to be closely fitted, which will be shown next.

The details of the model are given in [7]. A schematic of the physical layout of the circuit is shown in figure 6.3. By applying Kirchhoff's laws and the usual AC circuit analysis of the circuit shown in figure 6.4, we obtain equations (6.1) and (6.2), which can be used to derive the time dependence of the I–V curve of the RTD:

$$\frac{dV(t)}{dt} = \frac{I(t) - f(V)}{C} \tag{6.1}$$

$$\frac{dI(t)}{dt} = \frac{V_b - RI(t) - V(t)}{L} \tag{6.2}$$

where $V_b = V_{DC} - V_{th}$, and V_{DC} is the direc current (DC) bias voltage, $I(t)$ is the current through R and L, and $V(t)$ is the voltage across the RTD; $f(V)$ is the highly nonlinear voltage-current characteristic of the RTD and is discussed in chapter 2.

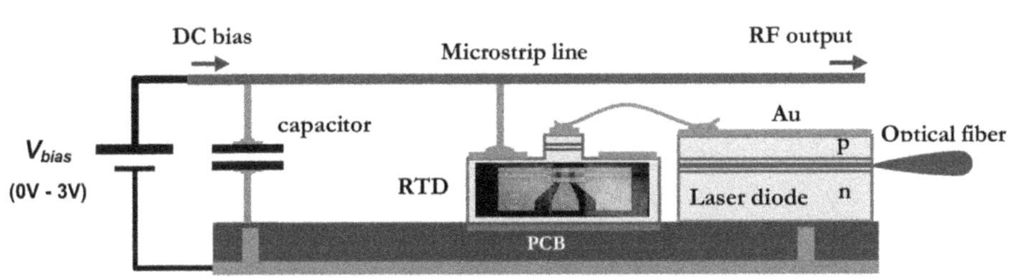

Figure 6.3. The electrical connections on a hybrid RTD-LD printed circuit board. © [2008] IEEE. Reprinted, with permission, from [7].

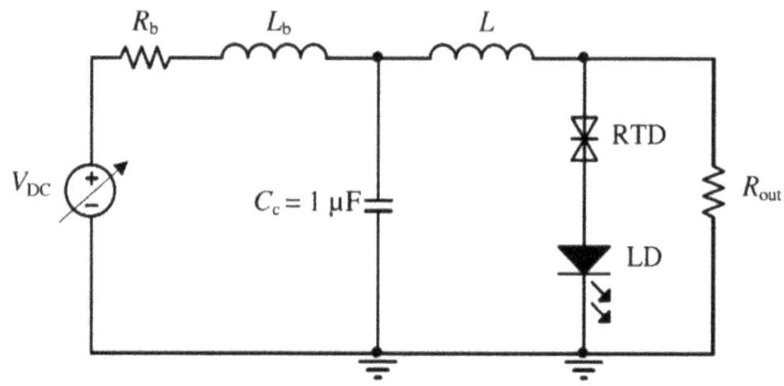

Figure 6.4. Schematic of the electrical circuit of the hybrid RTD-LD: R_b is the resistance associated with the bias circuit, L_b is the bias circuit inductance, L is the value of the inductance of the wire connecting the RTD and the LD, C_c is the capacitance and R_{out} is the output resistance. © [2008] IEEE. Reprinted, with permission, from [7].

Because of the NDC associated with $f(V)$, the circuit shown in figure 6.4 oscillates—the values of the external components in the circuit set the oscillation frequency.

The electrical current from the RTD is injected into the laser. In the following equations, N is the electron density, and S is the photon density:

$$\frac{dN}{dt} = \frac{I(t)}{q\mathcal{V}} - \frac{N}{\tau} - g_0(N - N_0)\frac{S}{1 + \epsilon S} \tag{6.3}$$

$$\frac{dS}{dt} = g_0(N - N_0)\frac{S}{1 + \epsilon S} - \frac{S}{\tau_p} + \frac{\beta N}{\tau} \tag{6.4}$$

$I(t)$ is the current through the laser equation (6.2); $q = | 1.6 \times 10^{-19} |$ C is the electronic charge; τ is the lifetime of an electron in the active mode volume, $\mathcal{V}$; τ_p is the lifetime of a photon in the active mode volume; N_0 is the minimum electron density required to obtain net gain; g_0 is the gain coefficient; β is the spontaneous emission factor; and ϵ is the nonlinear gain compression factor. Equations (6.3) and (6.4) describe how the electrical current from the RTD, $I(t)$, is injected into the active region of the laser diode and converted into an optical output.

Solutions from equation (6.1) through to equation (6.4) are used to predict the optical output from the RTD-LD. A comparison of the model to the experimental data is given in figure 6.5.

The Liénard oscillator model also predicts that the RTD-LD will operate as an optoelectronic, voltage controlled oscillator (VCO) [7], i.e., the fundamental oscillation frequency will depend on the bias voltage, V_{DC}, applied to the RTD-LD. Figure 6.6 shows the VCO characteristics of the RTD-LD and the fit of the Liénard oscillator VCO model to the experimental data of the DC bias voltage dependent frequency of the RTD-LD. A low inductance version of the same device could be tuned to 1.8–2.15 GHz by adjusting the DC bias [7].

The nonlinear model of the RTD-LD oscillator is more complete than linear models. The linear models of oscillator circuits [9] are reasonably accurate in predicting oscillation frequencies but often fail to predict output powers or oscillation waveforms.

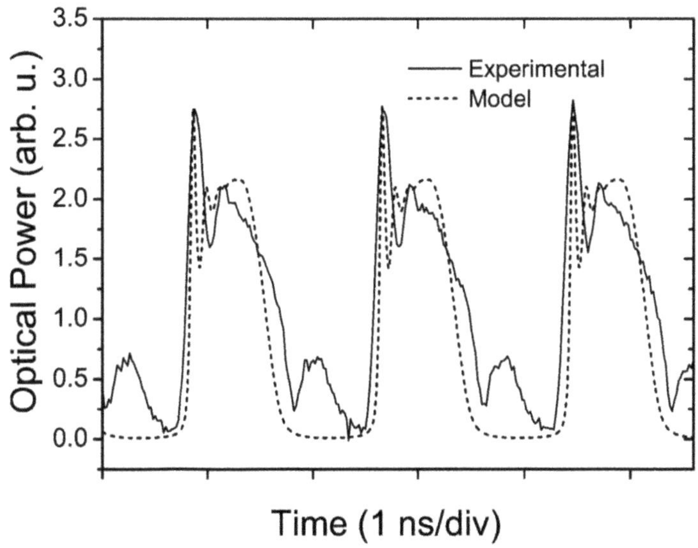

Figure 6.5. Experimentally measured output from the hybrid RTD-LD (solid line) and the modelled output (dashed line) using the Liénard oscillator model. The DC bias was set to 1.8 V and the fundamental frequency of oscillation is 560 MHz. © [2008] IEEE. Reprinted, with permission, from [7].

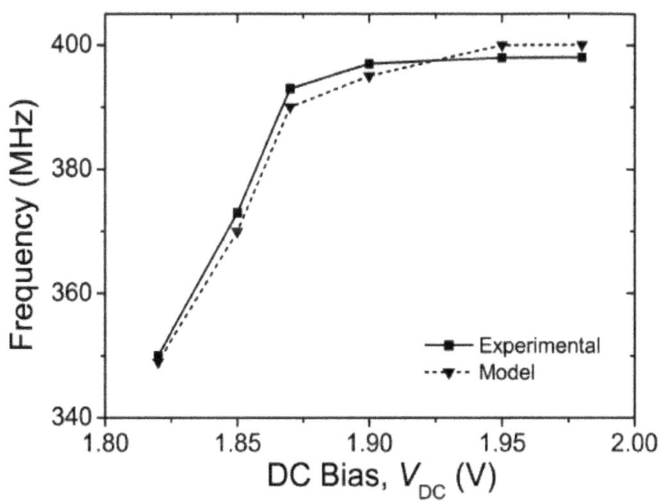

Figure 6.6. The fundamental frequency of the RTD-LD device versus the applied DC bias. The solid line is the experimental data and the dashed line is the curve fitted using the Liénard oscillator model. © [2008] IEEE. Reprinted, with permission, from [7].

6.3 Integrated RTD laser circuit

Previous work on monolithic integration of RTDs and AlGaAs/GaAs laser diodes focused on a laser emitting at ~850 nm [2]. There has also been work on using a multiple quantum well region as part of an RTD to obtain NDC and bistable operation [10]. Furthermore, there has been some work on observing the resonant tunnelling into quantum wells in a multiple quantum well laser but without the explicit addition of a double barrier RTD [11]. Tunnelling has been used to achieve large modulation bandwidths, 98 GHz, in a laser diode operating at 980 nm [12]. For a theoretical treatment of the properties of monolithic integration of a RTD and a quantum well laser, see [13]. This section discusses the first RTD monolithic integration with a laser diode emitting at optical communication wavelengths (around 1500 nm). The epitaxial layer, waveguide configuration and equivalent electrical circuit of an integrated RTD-LD designed to operate at optical communication wavelengths [3, 14] are shown in figures 6.7 and 6.8.

The RTD has three epitaxial layers: a 5-nm InGaAs quantum well between two strained (lattice mismatch) 2-nm AlAs barriers. The laser section has six 6.7 nm

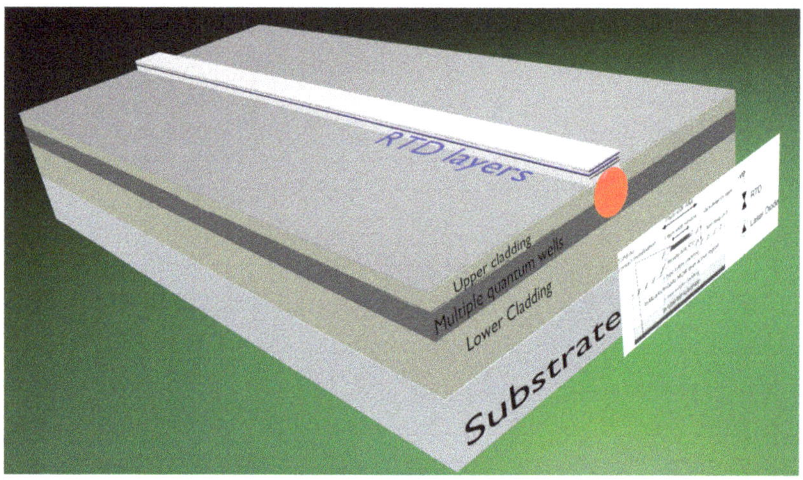

Figure 6.7. Schematic of an RTD-LD chip, which is a metal clad ridge waveguide laser with the RTD in the mesa that forms the ridge. The red dot indicates the region of the chip facet where light is emitted. A cross-section view is shown as an insert. For a better view of the cross section, see figure 6.8.

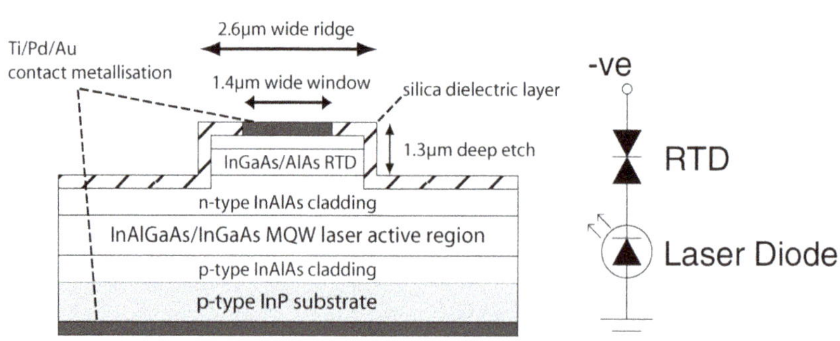

Figure 6.8. Cross section of the layout of the integrated ridge waveguide RTD-LD. The laser's resonator length is 500 μm, giving an RTD active area of 1300 μm^2. On the right-hand side is a schematic lumped element circuit representation of the device.© [2007] IEEE. Reprinted, with permission, from [3].

thick InGaAs quantum wells with 10 nm InGaAlAs barriers. Except for the AlAs RTD barrier layers, all of the epitaxial layers were lattice matched to InP, which was used as the substrate. The AlAs layers are thin enough not to disrupt the crystal growth because they are smaller than the critical layer thickness [15].

The RTD section was embedded in *n*-type material because the RTDs are conventionally operated with electrons rather than holes, plus it was necessary to place the RTD in the waveguide mesa to ensure that the current density through the device was sufficiently large to reach the NDR region of the RTD *I–V* curve with currents low enough not to cause thermal damage. Usually, semiconductor lasers are '*p*-side up', i.e., the waveguide channel layers furthest from the substrate are *p*-type doped. However, the RTD-LD with RTD layers on top of the waveguide in *n*-type material means that the RTD-LD is *p*-side down.

This RTD-LD operated at 1540 nm at room temperature, but the device could only be operated in pulse mode. To explore the device's continuous wave operation and clearly demonstrate the bistable operation, it was necessary to operate at cryogenic temperatures (130 K).

Figure 6.9 clearly shows that the device was bistable at cryogenic temperatures, i.e., the device has two stable states in terms of the DC bias applied: as the DC bias is increased, the device output flips on; and then as it is decreased, the device flips off at a lower voltage than the switch-on voltage that occurs as the voltage was increased. This means that the device has a hysteresis type of operation where the current state depends on past operation history. The difference between the on and off optical power output was ~30 dB.

The monolithic integration of a double barrier RTD and laser diode has been demonstrated at optical communications wavelengths and the RTD-LD device shows bistable operation, which could be particularly convenient for digital switching and optical memory operations [10].

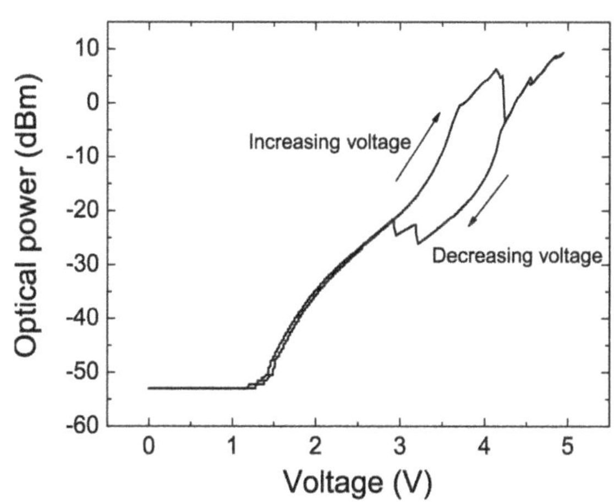

Figure 6.9. The optical output (in dBm) from the integrated RTD-LD operating 130 K versus the DC bias voltage shows clear bistability and hysteresis with >30 dB between the on and off states.© [2007] IEEE. Reprinted, with permission, from [3].

6.4 Integrated RTD light emitting diode (RTD-LED)

State-of-the-art micro- and nanophotonic devices rely on a *p–i–n* type architecture (as described in the previous section), which may result in additional electrical contact resistance and increased optical absorption losses. Developing monolithic integrated *n*-type unipolar (electron-transporting) light sources without *p*-type doping could substantially simplify the heterostructure band engineering that is typically required in *p–i–n* devices and lead to a lower cost fabrication.

A few studies have recently suggested that electroluminescence (EL) from *n*-type double barrier quantum well (DBQW) RTDs, namely GaN/AlN [16], can be achieved either due to impact ionisation or interband tunnelling effects as a result of the large electric field across the highly depleted DBQW under applied bias. However, thus far, EL from III–V DBQW materials has remained elusive and has only been reported at cryogenic temperatures [17], from the collector contact regions [18] or for *p–i–n* architectures [19] (as shown in the previous section). In another recent study [20], band-edge EL at ~1631 nm was observed in the emitter-doped side of a unipolar *n–i–n* InGaAs DBQW device with $\sim 2 \times 10^{-5}$ external quantum efficiency, which was enabled by hole generation due to both Zener interband tunnelling and impact ionisation mechanisms. However, thus far, room-temperature EL from the intrinsic active DBQW region, similarly as occurring in conventional *p–i–n* devices, remains challenging.

In this section, we cover the realisation of pronounced room-temperature radiative spontaneous light emission at ~806 nm (DBQW region emission) and at ~860 nm (near the band-edge transition) from a unipolar *n–i–n* RTD micro-light-emitting device (microRTD-LED) consisting of an AlAs/GaAs/AlAs DBQW. This device can potentially be scaled to sub-μm sizes avoiding high electrical contact resistance and optical absorption losses resulting from *p*-type doped materials. This paves the way for a new class of *n*-type optoelectronic device, such as micro- and nano-light sources.

6.4.1 Design and fabrication of unipolar microRTD-LEDs

The epilayer semiconductor layer stack of the *n*-type GaAs unipolar microRTD-LED grown by molecular beam epitaxy is shown schematically in figure 6.10(a), which consists of an unintentionally doped AlAs/GaAs/AlAs DBQW. The incorporation of thick AlAs barriers (3.1 nm) enables high fields ($>10^5$ V cm^{-1}) without excessive current flow and a significant charge build-up. The DBQW was surrounded by *n*-type lightly doped AlGaAs cladding layers, allowing high carrier injection efficiency and light confinement in the DBQW active region. Finally, the epilayer contained n^+-GaAs 2×10^{18} cm^{-3} emitter and collector layers. A thin (20 nm) top contact layer was chosen for efficient light extraction. Figure 6.10(b) shows the conduction and valence band (CB and VB) profiles under applied voltage. Under applied bias voltage, electrons that flow and build up in the quantum well recombine (radiative transitions $h\upsilon_{E1}$ and $h\upsilon_{LH1}$) with minority holes created due to impact ionisation or Zener interband tunnelling processes. Instead of injection from a *p*-contact, holes are created exactly at the DBQW region where carrier

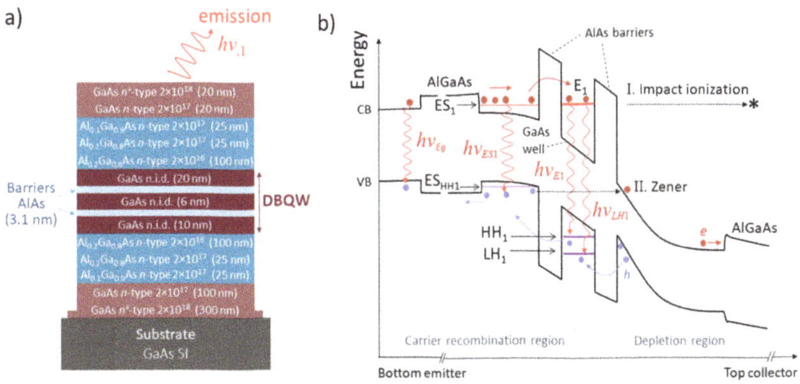

Figure 6.10. Semiconductor heterostructure of the unipolar microRTD-LED. (a) Semiconductor epilayer stack on a semi-insulating GaAs substrate showing the intrinsic double barrier quantum well (DBQW) active region. (b) Schematic of the energy conduction (CB) and valence band (VB) diagram under applied voltage, along with the flow of charge (electrons, *e*, and holes, *h*) that enable light emission (red downward arrows) in the carrier recombination region. This diagram shows a qualitative representation of the two main processes of electron–hole generation: I. impact ionisation and II. Zener (interband) tunnelling (HH: heavy hole, LH: light hole, ES: emitter spacer). Reproduced from [21]. CC BY 4.0.

recombination can occur, avoiding high electrical contact resistance and optical absorption losses resulting from *p*-doped materials. The schematic in figure 6.10(b) also depicts the spontaneous recombination processes that can occur in the spacer GaAs layer ($h\upsilon_{ES1}$) and in the *n*- and n^+-doped GaAs layer ($h\upsilon_{Eg}$) on the emitter side (and possibly on the collector—not shown).

The microRTD-LEDs were fabricated by defining micropillars ranging from 5 to 9 μm by electron beam lithography and etched down to ~550 nm by wet etching using a solution of H_2SO_4:H_2O_2:H_2O (1:2:20). The micropillars were passivated using a layer (~650 nm thickness) of SiO_2. The definition of the bottom contacts, via openings and top contacts employed optical lithography using direct laser writing. The top and bottom contacts were formed using Ge/Ni/Au metal alloys, followed by rapid thermal annealing. A scanning electron microscope (SEM) image of a representative fabricated microRTD-LED device with ~9 um diameter pillar is shown in figure 6.11.

6.4.2 Room-temperature electroluminescence

The EL was measured at room temperature with a ground–signal–ground (GSG) coplanar waveguide electrical probe connecting the devices. The electrical DC bias was provided with a source meter connected to a bias-T. A fibre-based micro-electroluminescence (μEL) system was used to collect the EL. The EL from microRTD-LEDs was collected by a multimode lensed fibre (spot size ~25 μm) connected to a fibre-coupled spectrometer to acquire the optical spectra in the near infrared region, or connected to a calibrated optical power meter for optical power measurements.

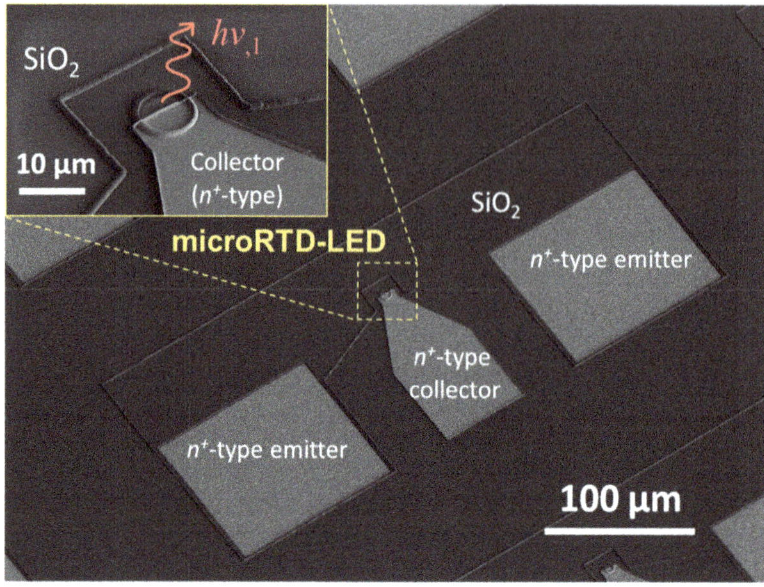

Figure 6.11. Light-emitting unipolar microRTD-LED device. SEM image of a fabricated unipolar microRTD-LED device with a micropillar of ~9 µm diameter. The inset shows a magnified image of the top micropillar half coated with the metal contact (n^+-type collector), enabling light out-coupling. Reproduced from [21]. CC BY 4.0.

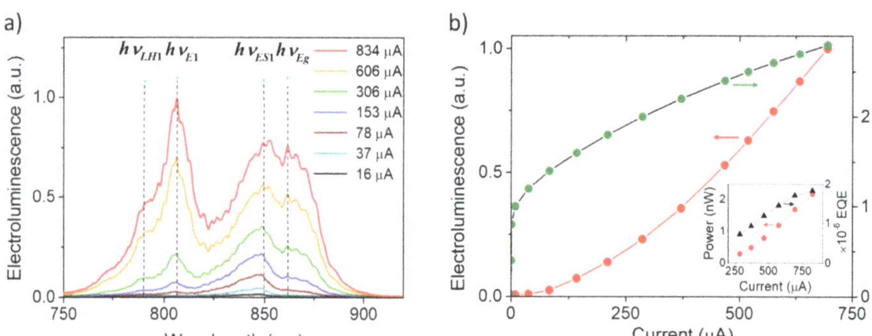

Figure 6.12. EL at room temperature. (a) EL spectra as a function of the pumping current (voltage ranging from 2 to 3 V) for a 7 µm wide pillar microRTD-LED. On top are marked the emission peaks $h\nu_{\mathrm{E1}}$, $h\nu_{\mathrm{LH1}}$, $h\nu_{\mathrm{ES1}}$ and $h\nu_{\mathrm{Eg}}$ (see figure 6.10(b)) (b) Light–current–voltage (L–I–V) characteristics of a 5 µm wide pillar unipolar microRTD-LED. The inset shows the collected optical power output (P_{out}) from the microRTD-LED as a function of the current, and the external quantum efficiency. Reproduced from [21], CC BY 4.0.

Figure 6.12(a) shows a typical EL spectrum as a function of the bias current for a microRTD-LED with a 7 µm pillar diameter. As a result of the high field achieved across the depletion region, the principal mechanisms related to hole generation include interband tunnelling and impact ionisation. However, given the estimated electric field strength for the experimental voltage utilised (<400 kV.cm^{-1} up to 3 V), we attribute the primary mechanism for hole generation of these devices to impact ionisation. As a result, we observe four EL emission signatures, see figure 6.12(a), which correspond to the interband transitions of the semiconductor layer stack, see figure 6.10(b): (i) a pronounced peak ($h\nu_{\mathrm{E1}} \sim 806$ nm) corresponding to the transition of an electron from the first energy level (E_1) in the conduction band of the DBQW to the heavy hole level (HH_1) in the VB; (ii) a shoulder emission ($h\nu_{\mathrm{LH1}} \sim 790$ nm), which corresponds to the transition of an electron from E_1 to the first energy level in the VB for light holes (LH_1); (iii) a peak emission ($h\nu_{\mathrm{ES1}} \sim 850$ nm) corresponding to the emitter spacer quantum well (ESQW) recombination of holes that are transported by resonant tunnelling to the emitter side with available electrons injected in the emitter spacer; and (iv) a broad peak peaking at $h\nu_{\mathrm{Eg}} \sim 861$ nm, which corresponds to the expected band-edge transition (E_g) at room-temperature from the GaAs bottom doped layers in the emitter side.

Figure 6.12(b) shows the measured light–current–voltage (L–I–V) characteristics of the unipolar microRTD-LED. The I–V has a diode-like turn-on voltage ~1 V, and the L–I increases monotonically within the measured current range. Because of the small contact area of the devices considered here, a large current density (up to 3 kA cm^{-2}) is obtained. The devices have a high series resistance (~2.47 kΩ), which is partly attributed to the micropillar device's small contact area and the contribution from the AlGaAs layers. NDR is also expected because the devices are built on a DBQW design. However, non-resonant current pathways seem to dominate. Thus, no clear observation of reproducible NDR was measured. As discussed next, signatures of NDR can be observed for larger area devices with improved fabrication methods.

6.4.3 Toward negative differential resistance

To achieve NDR from the I–V characteristic, an essential property of DBQW devices, from the same epilayer stack, see figure 6.10(a), another fabrication run was implemented to obtain a larger area ($d \sim 20$ μm) microRTD-LED. Here, dry etching was used and an improved passivation method was implemented employing Si_xN_y deposition by low-frequency PECVD [22]. From these fabricated microRTD-LEDs, reproducible signatures of NDR in the I–V characteristics were observed, see figure 6.13(a). Panel (b) in figure 6.13 shows the ratio of the maximum counts recorded from the quantum well (L_{QW}) to the emission from the GaAs emitter regions (L_{Bulk}) as a function of the bias voltage. We note that the overall EL spectra monotonously increase with the bias voltage, as shown in figure 6.13(b) at the peak and the valley voltage. However, the ratio $L_{\mathrm{QW}}/L_{\mathrm{Bulk}}$ exhibits a nonlinear behaviour. First, $L_{\mathrm{QW}}/L_{\mathrm{Bulk}}$ increases well above one until just before V_{p}, indicating that before the peak voltage (maximum overlap of the QW resonance with the Fermi energy distribution) the L_{QW} emission dominates. Then, the $L_{\mathrm{QW}}/L_{\mathrm{Bulk}}$ ratio starts to decrease as the bias increases, indicating that the L_{Bulk} emission on the emitter side becomes predominant. Interestingly, $L_{\mathrm{QW}}/L_{\mathrm{Bulk}} \sim 1$ when the current value at the V_{c} voltage is the same as the peak current, I_{p}. We attribute this nonlinear behaviour to the interplay between the resonant tunnelling and the non-resonant current

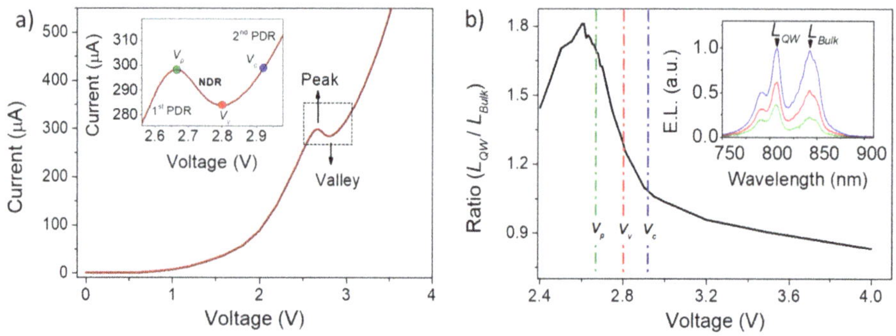

Figure 6.13. NDR and EL. (a) Current–voltage (I–V) characteristic of a unipolar microRTD-LED (d ~ 20 μm) showing the regions of NDR and positive differential conductance (PDR). The inset shows a magnified plot of the NDR region. The peak voltage, V_p, valley voltage, V_V, and V_c (bias current point identical to I_p) are marked as green, red and blue dots, respectively. (b) The ratio of the maximum counts in the quantum well (L_{QW}) to the GaAs emitter regions (L_{Bulk}), as a function of the bias voltage. The dashed vertical lines mark the V_p (green), V_V (red) and V_c (blue). The inset shows shown the EL spectra in the peak (green), valley (red) and V_c (blue) bias voltage conditions. Reproduced from [21]. CC BY 4.0.

components. Considering the range of applied bias within the NDR, we can assume a constant hole generation rate and that, in principle, holes have several channels to be transported either to the QW and to the emitter. However, the electron concentration in the quantum well is controlled by the applied voltage across the DBQW. As a result, beyond the peak voltage, an increase in the bias voltage leads to a reduction of the electron concentration in the well, resulting in a reduction of the QW EL, and explaining the reduction of the L_{QW}/L_{Bulk} ratio. In fact, this reduction should start slightly before the current peak (I_p) is reached, as observed in figure 6.13(b). This happens because the resonant component of the current reaches its maximum slightly before the current peaks. After all, the resonant current drop is initially compensated by the steady increase of the non-resonant current, and only then does the resonant current reduction become significant enough to give rise to the NDR region.

The observation of NDR and EL control from a single unipolar microRTD-LED paves the way to further optimise the epilayer stack to achieve much higher peak-to-valley-current ratio values, which could support optical switching or optical oscillations in a single RTD-LED device without the need for additional electronic or optoelectronic components.

6.5 Conclusions

RTDs can be integrated with laser diodes to form RTD-LD devices. Two integration layouts have been discussed: (i) the hybrid integration, where there are two separate chips, the RTD and the laser diode chips, that are connected with a short wire (~2 mm); and (ii) the fully monolithic integrated layout, with the RTD forming part of the epitaxial layers that make up the top contact layers in a semiconductor laser chip.

When DC is biased in the NDC region, the hybrid RTD-LD optical power output oscillates at frequencies set by the value of the external components—these frequencies can be >1 GHz.

Although so far operating continuous wave only at cryogenic temperatures (~130 K), the monolithically integrated RTD-LD has emitted at wavelengths (1540 nm) in the C band of the optical communications window and has demonstrated bistable operation [3].

In chapters 7 and 8, we discuss how the large electrical gain of the RTD-PD (chapter 5) can produce enough electrical power to drive the hybrid RTD-LD to form part of an OEO that employs optical fibre as the energy storage part of the oscillator. The RTD-based OEO is a version of the OEO that uses a minimum number of components in a highly efficient, low noise and high Q oscillator operating at RF frequencies >1 GHz.

Finally, we covered the first observation of pronounced room-temperature radiative spontaneous light emission at ~806 nm from the active DBQW material of a unipolar *n–i–n* microLED. Since highly doped *p*-type layers are not required, this novel microLED device based on RTD can be scaled down to sub-μm sizes, thus avoiding high electrical contact resistance and optical absorption losses resulting from *p*-type materials. This substantially simplifies the heterostructure band engineering that is typically required in *p–i–n* devices and could be extended to other III–V compound material systems, e.g., InGaAs, covering different wavelengths for a new class of *n*-type optoelectronic light-emitting devices.

References

[1] VanHoof C, DeNeve H, Mertens R, Romandic I, Goovaerts E and Borghs G 1997 Gigahertz modulation of tunneling-based GaAs light emitters *IEEE Photonics Technol. Lett.* **9** 1463–5

[2] Grave I, Kan S C, Griffel G, Wu S W, Saar A and Yariv A 1991 Monolithic integration of a resonant tunneling diode and a quantum-well semiconductor-laser *Appl. Phys. Lett.* **58** 110–2

[3] Slight T J and Ironside C N 2007 Investigation into the integration of a resonant tunnelling diode and an optical communications laser: model and experiment *IEEE J. Quantum Electron.* **43** 580–7

[4] Li Y J *et al* 2018 Electrically and optically bistable operation in an integration of a 1310 nm DFB Laser and a tunneling diode *Chin. Phys. Lett.* **35** 044202

[5] Chung S and Balkan N 2005 Fabry–Perot Gunn laser *Appl. Phys. Lett.* **86** 21111

[6] Feng M, Qiu J Y and Holonyak N 2018 Tunneling modulation of transistor lasers: theory and experiment *IEEE J. Quantum Electron.* **54** 2000514

[7] Slight T J, Romeira B, Wang L Q, Figueiredo J M L, Wasige E and Ironside C N 2008 A lienard oscillator resonant tunnelling diode-laser diode hybrid integrated circuit: model and experiment *IEEE J. Quantum Electron.* **44** 1158–63

[8] Guckenheimer J 1980 Dynamics of the Van-Der-Pol equation *IEEE Trans. Circuits Syst.* **27** 983–9

[9] Groszkowski J 1964 *Frequency of Self-Oscillations* (Oxford: Pergamon)

[10] Kawamura Y, Asai H, Matsuo S and Amano C 1992 InGaAs–InAlAs multiple quantum-well optical bistable devices using the resonant tunneling effect *IEEE J. Quantum Electron.* **28** 308–14
[11] Kawamura Y, Wakita K and Mikami O 1988 InGaAs/InAlAs bistable multiple quantum well lasers with large on-off ratio using the resonant tunneling effect *Appl. Phys. Lett.* **53** 1462–4
[12] Zhang X, GutierrezAitken A, Klotzkin D, Bhattacharya P, Caneau C and Bhat R 1997 0.98-micron multiple-quantum-well tunneling injection laser with 98-GHz intrinsic modulation bandwidth *IEEE J. Sel. Top. Quantum Electron.* **3** 309–14
[13] Sheng H Y and Chua S J 1994 Properties of monolithic integration of a resonant-tunneling diode and a quantum-well laser *Superlattices Microstruct.* **16** 157–60
[14] Slight T J 2006 Integration of a resonant tunnelling diode and an optical communications laser *PhD Thesis* (Department of Electronics and Electrical Engineering, University of Glasgow)
[15] Asahi H and Horikoshi Y (ed) 2019 *Molecular Beam Epitaxy: Materials and Applications for Electronics and Optoelectronics* (New York: Wiley)
[16] Growden T A, Zhang W D, Brown E R, Storm D F, Meyer D J and Berger P R 2018 Near-UV electroluminescence in unipolar-doped, bipolar-tunneling GaN/AlN heterostructures *Light-Sci. Appl.* **7** e17150
[17] White C R H, Skolnick M S, Eaves L and Leadbeater M L 1991 Electroluminescence and impact ionisation phenomena in a double barrier resonant tunneling diode structure *Appl. Phys. Lett.* **58** 1164–6
[18] Hartmann F *et al* 2017 Temperature tuning from direct to inverted bistable electro-luminescence in resonant tunneling diodes *J. Appl. Phys.* **122** 154502
[19] Buhmann H, Mansouri L, Wang J, Beton P H, Eaves L and Henini M 1994 High-efficiency submicron light-emitting resonant-tunneling diodes *Appl. Phys. Lett.* **65** 3332–4
[20] Brown E R, Zhang W D, Growden T A, Fakhimi P and Berger P R 2021 Electroluminescence in unipolar-doped In0.53Ga0.47As/AlAs resonant-tunneling diodes: a competition between interband tunneling and impact ionization *Phys. Rev. Appl.* **16** 054008
[21] Jacob B, Camarneiro F, Borme J, Figueiredo J M L, Nieder J B and Romeira B 2023 Room-temperature electroluminescence and light detection from III–V unipolar microLEDs without p-type doping *Optica* **10** 528–37
[22] Jacob B, Camarneiro F, Borme J, Bondarchuk O, Nieder J B and Romeira B 2022 Surface passivation of III–V GaAs nanopillars by low-frequency plasma deposition of silicon nitride for active nanophotonic devices *Acs Appl. Electron. Mater.* **4** 3399–410

IOP Publishing

Resonant Tunnelling Diode Photonics Devices and Applications (Second Edition)

Charlie Ironside, Bruno Romeira and José Figueiredo

Chapter 7

Nonlinear dynamics of RTD oscillators

Chapter 7 explores the nonlinear dynamics of oscillators that have resonant tunnelling diodes (RTDs) as the key component and how RTD oscillators can be controlled by either electrical or optical signals. Compared to linear systems theory, nonlinear dynamics provides a much more comprehensive understanding of the behaviour of RTD oscillators. The key dynamic scenarios covered in this chapter include injection locking, period-adding bifurcation, quasi-periodic route to chaos, and introduces neuromorphic applications using neuron-like excitable spiking dynamical phenomena.

7.1 Introduction

There is a substantial interest in developing novel transmitters and receivers to meet the future needs of communication and computing systems. Among the various approaches, the implementation of electro-optical and all-optical solutions that take advantage of the unique nonlinearities of semiconductor active devices, e.g., optoelectronic oscillators, laser diodes and modulators, has attracted considerable attention in the last few decades. The key applications of nonlinear photonic devices include microwave-photonic systems for signal processing [1, 2], secure optical chaos communications [3, 4] and more recently neuromorphic (brain-inspired) photonic applications [5].

As discussed in chapters 5 and 6, the recent advances on resonant tunnelling diode-laser diode (RTD-LD) resonant tunnelling diode-light-emitting diode (RTD-LED), and resonant tunnelling diode-photodetector (RTD-PD) modules have established a new-type of RTD-based photonic circuit that has led to significant breakthroughs, including the demonstration of dynamical regimes such as voltage-controlled autonomous relaxation oscillations in both electrical and optical domains. In this chapter, we cover the nonlinear dynamic properties of non-autonomous RTD photonic oscillators, i.e., subjected to external perturbations.

doi:10.1088/978-0-7503-5714-2ch7

The key dynamic scenarios covered in this chapter include injection locking, period-adding bifurcation, quasi-periodic route to chaos and excitable, i.e., neuron-like, dynamics. In chapter 9, we cover in detail the excitable neuron-like properties of RTD photonic devices for neuromorphic computing applications.

7.2 Photonic synchronisation and chaos

7.2.1 Injection locking

Injection locking is a well-known phenomenon that is discussed in standard textbooks on nonlinear oscillators [6, 7], under headings such as synchronisation, phase locking, frequency entrainment or forced oscillation. According to the oscillator model developed by Adler and described in a seminal paper in [8], an oscillator can be locked in frequency by an external signal that is injected into the oscillator. The mechanism of the locking process depends upon the initial frequency difference between the oscillator, the relative amplitude between the injected signal and the oscillator, and the respective circuit parameters.

In this subsection, we investigate the injection locking characteristics of RTD-LD oscillators to incoming radio frequency (RF) signals. In figure 7.1(a), a photograph of a typical RTD-LD hybrid circuit module mounted on a printed circuit board is shown (see chapter 6 for more details on the RTD-LD circuit and its free-running autonomous operation). The respective diagram showing both the equivalent circuit and the experimental setup is depicted in figure 7.1(b). For the injection locking experiment, an electrical signal is injected into the electrical input port using a bias-tee and the output signal is coupled from the laser diode's optical output using a lensed fibre to a photodetector. The RF spectra of the free-running and injection-locked RTD-LD outputs are typically measured using a fast spectrum analyser or oscilloscope.

In figure 7.2(a), a typical spectrum (black dashed–dotted line) of the RTD-LD's free-running oscillations is presented (in this example, the oscillation frequency is set to around 1.24 GHz). The corresponding single side band (SSB) phase noise measurements are shown in figure 7.2(b). The photo-detected free-running signal from the laser's output shows a low-quality phase noise, and a large frequency drift

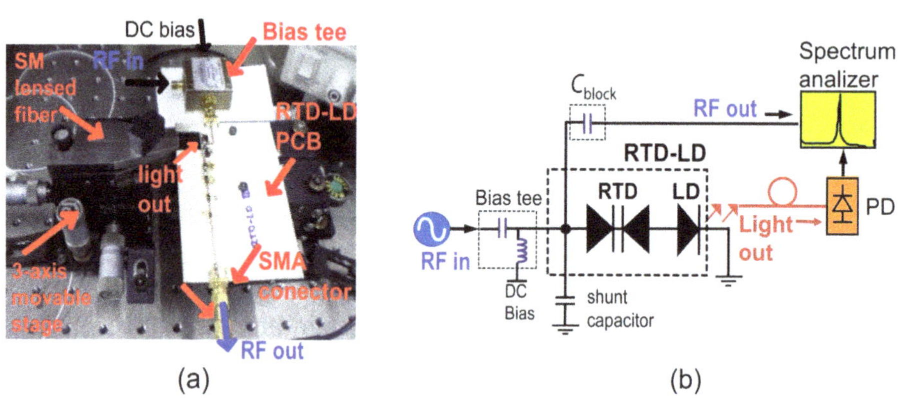

Figure 7.1. RTD-LD hybrid integrated circuit. (a) Photograph of the RTD-LD circuit. (b) Circuit schematic diagram of the experimental setup.

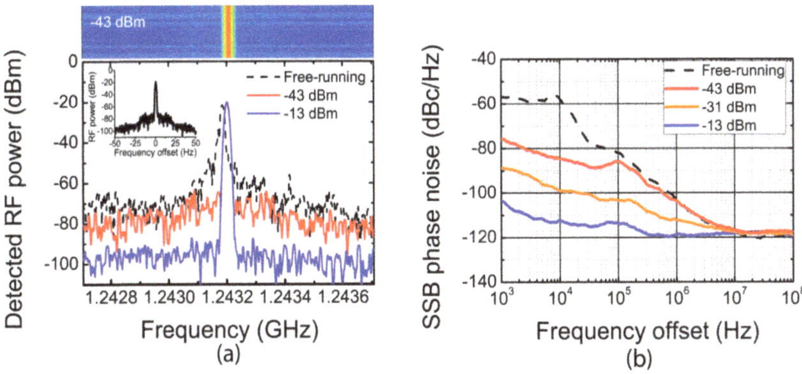

Figure 7.2. Injection locking. (a) Free-running and injection-locked RF spectra of the photo-detected RTD-LD outputs (1 MHz span and 10 kHz RBW). The inset shows the injection-locked spectrum in the case of locking with an incoming RF signal with −13 dBm power level (100 Hz span and 1 Hz RBW). The top inset shows a spectrogram that represents the evolution of spectral density of the injection-locked signal recorded over approximately 60 s (the vertical axis represents time). (b) SSB phase noise measurements for both free-running oscillations and injection-locked signals.

and instability. The corresponding SSB power density is −81.96 dBc Hz^{-1} at 100 kHz offset of the carrier centre frequency.

When an RF signal frequency is injected at a frequency that is sufficiently close to the free-running oscillation frequency (even using extremely low power levels, in this case −43 dBm), the RTD electronic oscillation is stabilised and pulled to the same frequency as the incoming RF signal. The corresponding modulated laser output signal is also injection-locked and highly-stable with a corresponding SSB phase noise of −85.86 dBc Hz^{-1} at 100 kHz offset of the centre frequency. We note that, in the case of the injected power being lower than the free-running signal, the amplitude of the injection-locked RTD-LD is similar to the amplitude of the free-running signal. The inset on the top of figure 7.2(a) shows the spectrum surveillance confirming the spectral stability of the locked signal. The pull or locking range Δf was also measured. For this level of injected power, the locking range was around 300 kHz. For injected power levels much lower than the power of the RTD-LD free-running oscillator, the locking range was found to be approximately proportional to $\sqrt{P_i/P_0}$, where P_i is the injected power and P_0 is the RF output power of the RTD-LD free-running signal [9]. This behaviour follows Adler's theory [8] and the injection locking mechanisms found in other conventional negative-resistance oscillators operating in the small-signal limit.

We note that increasing the output power of the RF synthesiser further decreases the phase noise level of the reference signal. As the noise level of the input is decreased, the phase noise of the injection-locked signal decreases accordingly. This is demonstrated in the spectrum (blue trace) of figure 7.2(a) using a power level of the incoming RF signal of −13 dBm. Figure 7.2(b) shows the corresponding low phase noise with −113.69 dBc Hz^{-1} phase noise at 100 kHz carrier offset. The injection-locked RTD-LD oscillator follows the noise characteristics of the input signal, limited only by the instrument's noise floor. The inset of figure 7.2(a) shows the same injection-locked signal but using a 1 Hz resolution bandwidth (note the different frequency scales), confirming that the output spectrum collapses to match that of the reference synthesiser. We achieve a suppression of more than 30 dB when compared with the free-running oscillating signal. The corresponding locking range has increased to 30.5 MHz, corresponding to approximately 2.45% of the oscillation frequency.

7.2.2 Period-adding bifurcation

To map the nonlinear dynamical behaviour of the RTD-LD photonic circuit subjected to external injection, we have characterised in detail the electro-photonic outputs in both time and Fourier domains for a wide range of voltage amplitude, V_{AC}, and injected frequency, $f_{in,}$ values. The observed dynamic trajectories were plotted in the voltage–light (V–S) phase space, where V represents the voltage across the RTD-LD (electrical output), and S is the laser diode's optical output. In the free-running case, the trajectories in the V–S phase space correspond to a simple attractor, also known as limit cycle, as shown in figure 7.3(a). The corresponding free-running frequency was $f_0 \sim 0.95$ GHz. We note in this example the circuit analysed was oscillating at frequencies, f_0, ranging from 0.5 GHz to 1 GHz, controlled by the direct current (DC) voltage [10]. The corresponding spectra of both the free-running electrical and optically modulated oscillations are presented in figure 7.3(b), showing an harmonic content up to the 10th harmonic.

Following the injection locking phenomena discussed in the previous subsection, a more complex circuit behaviour known as period-adding bifurcation dynamics [11] can be obtained when the free-running oscillations are perturbed by injected RF signals at frequencies that do not correspond to the fundamental frequency of the oscillator. To map these dynamic regimes, whenever a stable closed trajectory was found in the V–S phase space the system, we considered this a phase-locked regime. We also considered a very weak perturbation regime, $V_{AC} \ll \Delta V_{NDC}$, where ΔV_{NDC} is the negative differential conductance (NDC) voltage width (~250 mV for the circuits tested). In these conditions, the free-running and injected frequencies, f_0 and f_{in}, respectively, satisfy the relation $f_{in}/f_0 \rightarrow n/m$, where n and m are natural numbers. The ratios n/m are known as winding numbers [6, 7].

Figures 7.3(c)–(j) present examples of RTD-LD synchronised states using a sinusoidal input voltage amplitude of 300 mV and scanning the injected frequency from 0.3 GHz to 3.5 GHz. On the left-hand side are shown the V–S phase space diagrams and the corresponding spectra are shown on the right-hand side. Figures 7.3(c)–(d) show locking at a frequency that is half of the free-running oscillation (1:2). Figures 7.3(e)–(f) show locking at signals with a frequency close to the free-running oscillation (1:1)—fundamental locking; Figures 7.3(g)–(h) and 7.1(i)–(j) illustrate two examples of harmonic injection locking, 2:1 and 3:1, respectively, where the injected frequencies are close to RTD-LD's first and second harmonic frequencies, respectively.

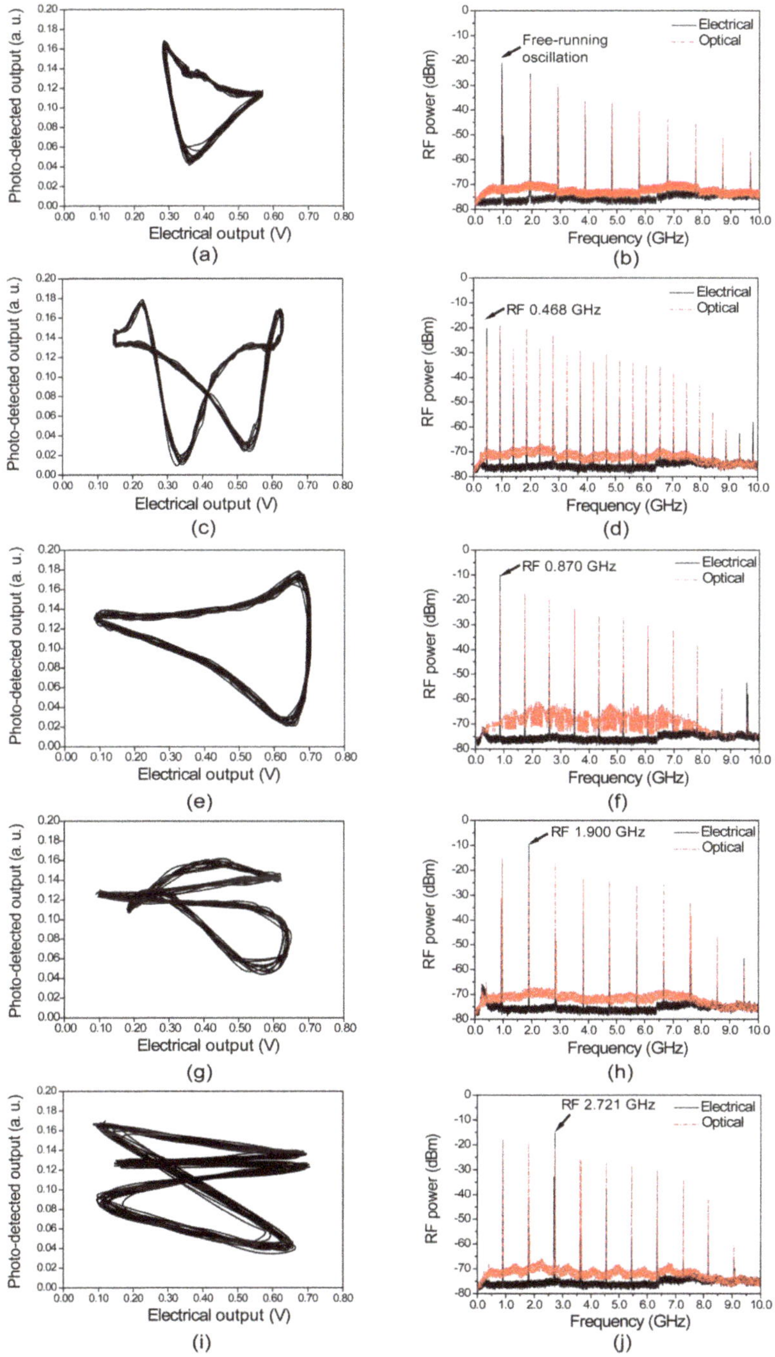

Figure 7.3. Experimental electro-photonic limit cycle regimes in the V–S phase space (left-hand panels) and corresponding spectra (right-hand panels) for the following stable dynamic states: (a)–(b) free-running; and locking: (c)–(d) 1:2, (e)–(f) 1:1, (g)–(h) 2:1, and (i)–(j) 3:1.

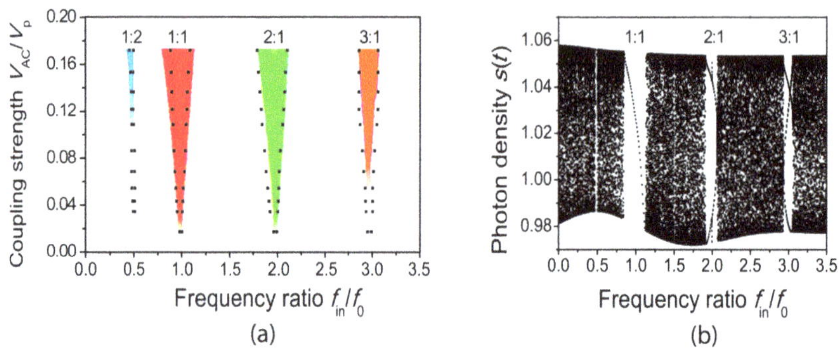

Figure 7.4. Period-adding bifurcation dynamics. (a) RTD-LD oscillator frequency-locking structure showing the electro-photonics Arnold tongues. The results from the experiment (dotted lines) are compared with theory (coloured tongues). (b) Numerical simulated bifurcation map of photon density, $s(t)$, as a function of frequency for the case of electrical injection of a periodic signal at a fixed low amplitude.

The overall circuit dynamic response to perturbations was mapped in a two-dimensional Arnold tongues diagram and is shown in figure 7.4(a). The tongues represented in the map correspond to the regions of locking to the driving signals presented in figures 7.3(c)–(j). The Liénard's numerical model, see section 7.3.2, agrees qualitatively well with the experimental tongues. The y-axis represents the perturbation strength defined as the ratio of the injected voltage amplitude V_{AC} and RTD-LD peak voltage V_p (around 2.90 V), V_{AC}/V_p; the x-axis corresponds to the frequency ratio f_{in}/f_0. The injected amplitude and frequency ranged from 50 to 500 mV and from 0.3 to 3.5 GHz, respectively. Figure 7.4(b) presents the corresponding simulated bifurcation map of the photon density, $s(t)$, showing the period-adding bifurcation dynamics sequence between intervals of unlocked signals. The results presented in figure 7.4 demonstrate that the RTD-LD circuit follows period-adding bifurcation dynamics. The modelling results confirm that under the regimes of operation analysed here, the RTD-LD circuit follows the RTD's nonlinearity in such a way that the laser dynamics follow exactly the dynamics of the RTD oscillator.

7.2.3 Quasi-periodic route to chaos

Outside the phase-locked regions, the circuit provides unsynchronised complex outputs that can be regarded as a result of the nonlinear mixing of the injected signal with the free-running oscillations. This results in quasi-periodic signals displaying a large frequency content, and chaotic signals with a broadband (GHz wide) spectrum. Figure 7.5(a) shows an example of a quasi-periodic trajectory in the unsynchronised region between the winding numbers 1:1 and 2:1. In contrast to previously synchronised orbits, the trajectories of quasi-periodic signals do not form a closed loop. Instead, the orbit circulates around a confined region known as quasi-periodic torus. Figure 7.5(b) shows the corresponding power spectra of V and S, showing a spike-like frequency distribution, both having the same frequency content. By increasing the nonlinearity by injecting a signal with a much larger

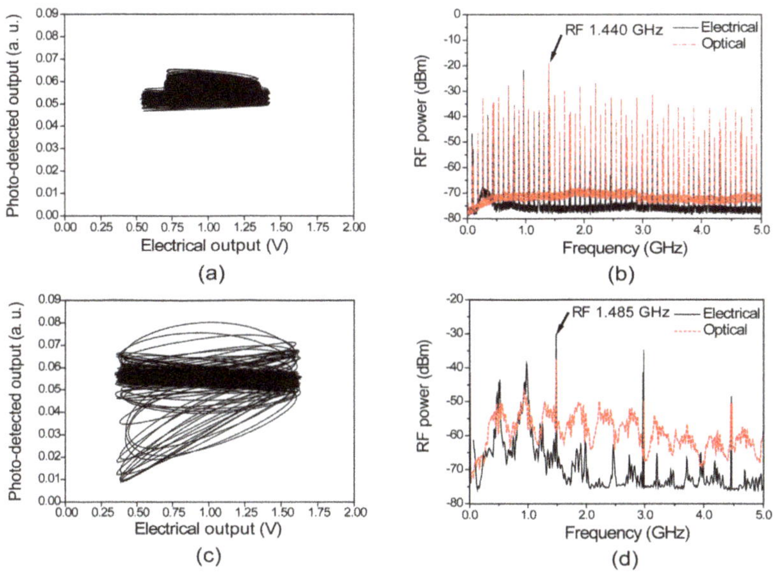

Figure 7.5. Experimental electro-photonic regimes in the *V–S* phase space (left-hand panels), and corresponding power spectra (right-hand panels) for (a)–(b) quasi-periodic, and (c)–(d) chaotic dynamic states. © [2009] IEEE. Reprinted, with permission, from [10].

amplitude than 300 mV, the circuit's dynamics evolve from synchronisation to quasi-periodicity and then to chaos, where chaotic burst signals begin to appear outside the torus as the frequency ratio f_{in}/f_0 converges to a certain irrational number. We observed these complex dynamics experimentally in a wide range of injection amplitudes. Figure 7.5(c) shows a typical chaotic trajectory for a driving amplitude of 790 mV. The corresponding spectra of the chaotic signal is displayed in figure 7.5(d). The sharp spike-like frequency peaks of quasi-periodic signals, figure 7.5(b), disappear in the transition to the chaotic regime, giving rise to signals with a broadband continuous frequency distribution. The peaks in figure 7.5(d) correspond to the injected RF signal and to minor traces of quasi-periodicity, indicating that the system's dynamics follow a quasi-periodic route to chaos. A significant increase of the noise level (up to 20 dB) is also observed, which is a key characteristic of chaotic signals.

The intrinsic instability and sensitivity to initial conditions of the chaotic signals in the RTD-LD photonic circuits was numerically investigated by evaluating the rate of divergence of nearby trajectories in the unsynchronised regions. Figure 7.6 presents a histogram of the largest Lyapunov exponent, λ_1, illustrating the frequency bands where the RTD-LD behaves chaotically, i.e., where $\lambda_1 > 0$. The largest chaotic band corresponds to f_{in}/f_0 around the golden number $(\sqrt{5} + 1)/2 \sim 1.62$ (also called the golden mean). The golden number, and other irrational numbers including the silver number, are commonly used in nonlinear dynamical studies [6]. The frequency range of chaotic operation and the chaos signatures, specifically the broadband spectrum shown in figure 7.5(d), indicate that the circuit is operating in the golden

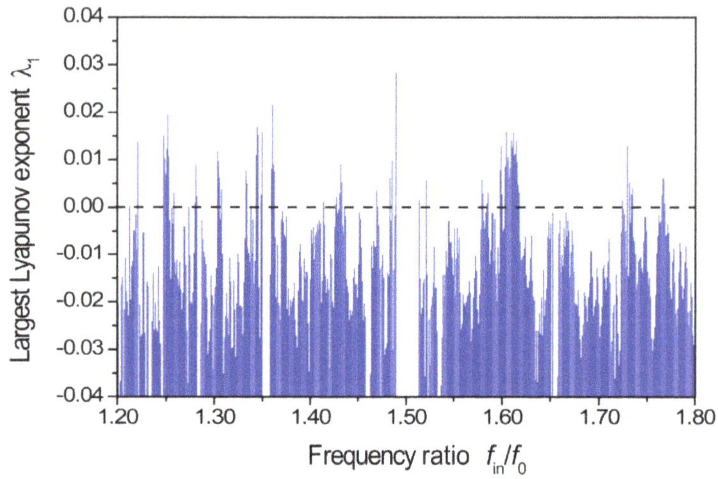

Figure 7.6. Histogram of the largest Lyapunov exponent for the unsynchronised region between 1:1 and 2:1 locking windows for the experimental conditions presented in figure 7.5. © [2009] IEEE. Reprinted, with permission, from [10].

ratio band [6] in the largest chaotic band shown in figure 7.6 because the intrinsic free-running frequency red shifts with the amplitude of the injected signal, leading to a value of f_{in}/f_0 close to the golden ratio. The analysis of figure 7.5(d) indicates that the free-running frequency (second peak in the spectrum) under this level of injection was shifted from 953 MHz to ~930 MHz, which gives $f_{\text{in}}/f_0 \sim 1.6$.

In summary, the process of achieving broadband chaotic optical sub-carriers in the laser output described here is physically different from the laser-based methods presented in [4, 12, 13]. The generation and conversion of chaotic currents into chaotic light intensity patterns discussed here depend solely on the RTD non-linearities. Since monolithic photonic integration leads not only to a considerable reduction in size, energy consumption and cost but also to an increase in speed and reliability, RTD-LD monolithic photonic circuits have potential as an alternative to the chaotic systems reported, for example in [13], to operate as the core of electro-photonics chaotic broadband emitters to encode data at the physical level.

7.3 Excitable spiking in neuromorphic photonic devices and systems

7.3.1 Excitability (neuron-like) dynamics

Excitability is a well-known nonlinear property [14] that regulates several biological process, such as nervous impulse propagation [15] or chemical reactions such as Belousov–Zhabotinsky systems [16]. Excitable systems provide an all-or-none response to external stimuli, i.e., the system responds linearly and with small amplitude to a stimuli that do not exceed a threshold, but otherwise produces a large pulse (whose shape is independent of the stimulus) followed by a lapse, known as the refractory time, during which the system does not respond to new stimuli. Excitable systems show potential to perform complex information processing tasks for applications in clock

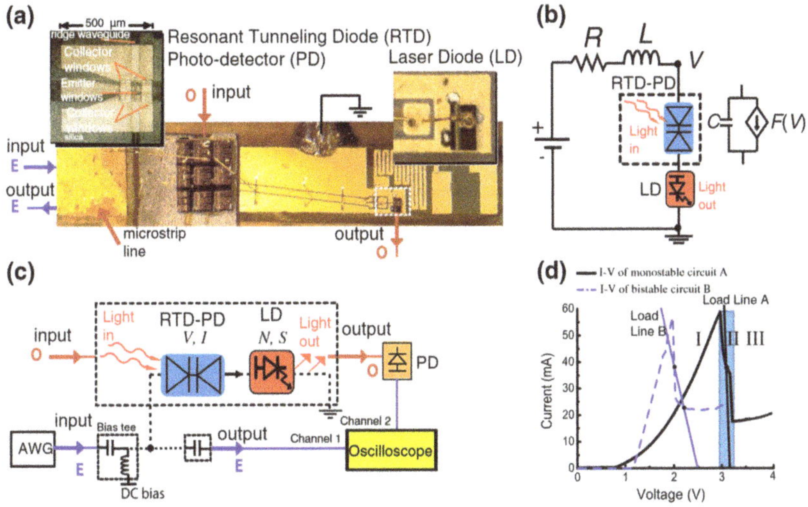

Figure 7.7. Neuromorphic photonic circuit. (a) Photograph of the hybrid optoelectronic integrated circuit. The inset shows the RTD and LD dies. (b) Corresponding lumped electrical circuit showing the equivalent inductance, *L*, resistance, *R*, and RTD-PD capacitance, *C*, and nonlinear *I–V*, *F*(*V*). (c) Schematic block diagram of the RTD in series with the LD forming the optoelectronic oscillator with quadruple electronic and optical inputs/outputs. (d) Experimental *I–V* curves of the monostable (solid) and bistable (dashed) circuits. Reprinted by permission from Springer Nature [25], Copyright (2014).

recovery and pulse reshaping functionalities in photonics [17]. However, the currently proposed systems operate at low speed and are bulky, which makes of the majority of them unsuitable for current information processing needs.

In this section, we investigate a high-speed neuromorphic photonic circuit using the RTD-LD photonic circuit configuration analysed in the previous sections. The typical circuit used to investigate the excitable dynamics is shown in figure 7.7. In chapter 9, we cover the excitable neuron-like properties of RTD photonic devices analysed here for neuromorphic computing applications.

The circuit displays an asymmetric DC current–voltage (*I–V*) curve with a pronounced negative differential conductance region, see figure 7.7(d). When DC biased in the NDC, it can operate as high-speed oscillators for THz generation [18]. RTD-based optoelectronic oscillator (OEOs) have recently been proposed for high-purity phase-locked oscillations employing either external injection locking, as analysed in the previous section of this chapter, or self-injection locking in a delay line feedback configuration [19], which will be analysed in chapter 8.

However, when DC biased outside the NDC, RTD-based photonic circuits can display new phenomena, including (i) bistable behaviour [20], and (ii) excitable behaviour as reported in [21], where optical pulse generation with refractory time responses in the order of the nanosecond were demonstrated. In this section, we cover this excitability phenomena characterised by noise activated dynamics when RTD-based circuits operate in either monostable or bistable regimes. In these regimes, we observe spikes, square pulses, bursting, and stochastic and coherence resonances (CRs) depending on the working conditions of the RTD photonic circuits. The experimental results are well described when considering the Liénard's oscillator nonlinear model that describes the RTD being stochastically activated by means of white Gaussian noise and driving the laser diode, which is described by the rate equations model, see chapter 6. The stochastically induced dynamics can be exploited in a variety of applications, including optoelectronic memories, or in artificial neural networks by coupling multiple excitable units.

7.3.2 Spiking and bursting dynamics

Figure 7.7(a) shows a photograph of the analysed circuit, and figure 7.7(b) presents the equivalent electrical circuit. We measured the response of the circuit to injected white Gaussian noise signals using an Agilent 33250A arbitrary waveform generator (AWG) by recording the electrical and optical outputs employing a 2 GHz oscilloscope, see figure 7.7(c).

Our circuit's operational quiescent point is defined as the intersection of the nonlinear I–V curve and the load line, whose slope is defined by the inverse of the circuit's resistance. As illustrated in figure 7.7(d), for low resistance, the load line is nearly vertical. There is only one intersection with the nonlinear characteristic function, allowing for operation in either a rest state regime, regions I and III, or a self-oscillatory regime, region II, depending on the bias conditions, whereas several (three) intersections may occur for high resistance.

As detailed in chapter 6, the equivalent photonic circuit is modelled by the single mode laser rate equations describing the laser diode's photon and carrier number density (S, N) dynamics coupled to a Liénard system accounting for the current and voltage across the RTD-LD. The model is summarised here as follows:

$$\dot{V} = \frac{1}{\mu}[I - f(V) - \chi\xi(t)] \tag{7.1}$$

$$\dot{I} = \mu[V_0 - \gamma I - V] \tag{7.2}$$

$$\dot{N} = \frac{1}{\tau_n}\left[\frac{1}{I_{th}} - N - \frac{N - \delta}{1 - \delta}(1 - \epsilon S)S\right] \tag{7.3}$$

$$\dot{S} = \frac{1}{\tau_p}\left[\frac{N - \delta}{1 - \delta}(1 - \epsilon S)S - S + N\beta\right] \tag{7.4}$$

Here, time is scaled by the LC resonance frequency $\omega_0 = (\sqrt{LC})^{-1}$, $f(V)$ describes the I–V curve, and $\mu = V_0/I_0\sqrt{C/L}$ and $\gamma = R(I_0/V_0)$ describe the corresponding equivalent circuit parameters (inductance, L, capacitance C, and resistance R), see figure 7.7(b). Finally, we incorporate the stochastic effects into our model by using an effective delta-correlated Gaussian white noise, $\xi(t)$, with zero mean and unit variance, where χ stands for the total noise strength.

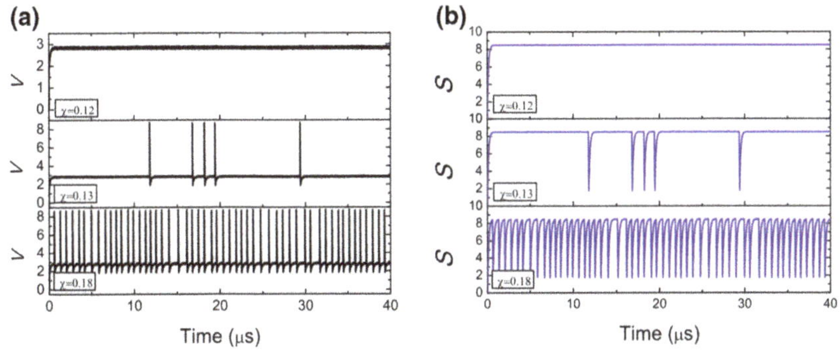

Figure 7.8. Simulated noise induced resonances in the RTD-PD-LD excitable system in the (a) electrical (V) and (b) optical (S) domains as a function of the noise strength. Reprinted by permission from Springer Nature [25], Copyright (2014).

When the RTD-LD works in monostable mode and is DC biased inside the NDC, region II, see figure 7.7(d), it drives the laser with voltage-controlled self-sustained current oscillations. When DC bias is applied in regions I and III, see figure 7.7(d), the RTD-LD responds to external perturbations by generating spikes in both the electrical and optical domains when the perturbation exceeds a predetermined threshold [21]. In this excitable regime, the system is very sensitive to noise, emitting fixed-form spikes or bursts of spikes depending on the DC bias point, and showing CR [14], i.e., a maximum of temporal regularity in the output for a finite noise level. Figure 7.8 displays a numerical example of spiking and CR behaviour in the RTD-LD due to noise perturbation and selecting the bias operation in region I, i.e., in an equilibrium state in the positive differential resistance. For an extremely low noise input level, $\chi = 0.12$, the RTD-LD output remains in its steady (rest) state. However, for a certain noise threshold perturbation, the RTD-LD can leave the rest state. We observe the output spiking in both electrical and laser outputs (V and S, respectively) when $\chi = 0.13$ (low noise). The peaks appear to be random and the time between excitations varies substantially. For a modest noise, $\chi = 0.18$, the spiking is rather regular, implying that the inter-spike intervals (ISIs) do not differ significantly, demonstrating CR behaviour.

The excitable pulse triggering mechanism is explained as follows. The RTD-LD can be considered as a slow–fast excitable system, with the fast variable being the voltage and the slow variable being the current. During the excitable orbit, the two fast stages correspond to a sudden peak and drop of the voltage, during which the current does not change appreciably. These two fast stages are interleaved by two slow stages in which the current evolves along the attracting nullcline defined by the $f(V)$ function, while the voltage follows adiabatically.

The excitable regimes predicted in the numerical simulations were verified experimentally. By adding a stochastic voltage generated by a Gaussian noise source with a cut-off frequency of 80 MHz to the DC bias in region I, the RTD-LD circuit A passes from emitting occasional spikes, see figure 7.9(a), plots (i) and (ii), to emission of an almost periodic pulse train, see figure 7.9(a), plot (iii) and (iv), as the

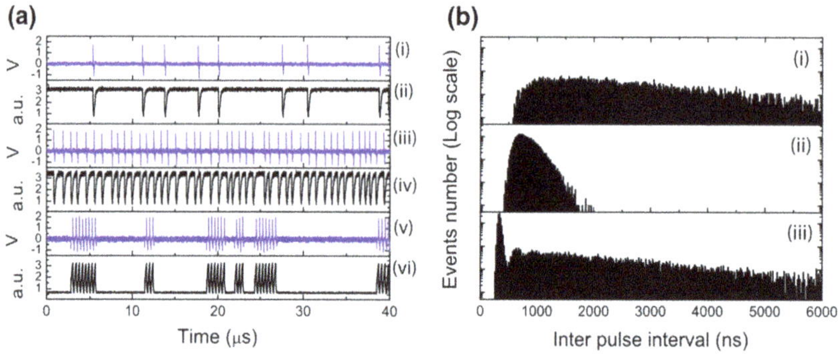

Figure 7.9. Experimental excitable spiking generation in the RTD-LD. (a) Experimental time traces of electrically noise induced neuron-like phenomena in both the electrical (V) and optical (a.u.) domains: (i)–(ii) random pulsing firing (V_{dc} = 2.85 V and V_{noise} = 180 mV); (iii–iv) CR behaviour (V_{dc} = 2.85 V and V_{noise} = 225 mV); and (v)–(vi) multi-pulsing bursting behaviour (V_{dc} = 3.36 V and V_{noise} = 250 mV). (b) Corresponding histograms of the ISI statistics of the laser output: (i) random pulsing firing; (ii) CR behaviour; and (iii) multi-pulsing bursting behaviour. Reprinted by permission from Springer Nature [25], Copyright (2014).

noise level increases. The upward electrical pulses have a full width half maximum (FWHM) of ~13 ns, while the laser diode's light output exhibits downward pulses with FWHM of ~200 ns. The LD emits pulses in response to current modulation and its response is determined by the carrier and photon lifetimes, yielding optical pulses longer than the current spikes. The photonic circuit has the potential to achieve sub-ns response [21], limited by the parasitics, e.g. the gold wires inductance, and by the laser relaxation oscillation in a monolithic circuit version. Further increasing the noise level above 225 mV destroys the periodicity of the signal, yet the shape of the pulse remains identical. Further analysis of the corresponding stochastic induced dynamics ISI series data, see figure 7.9(b), reveals the pulse statistics showing a typical exponential behaviour of a Kramer's escape process [14], displaced by the refractory time of the excitable orbits. Figure 7.9(b) (ii) shows the ISI of CR.

Similar behaviour is observed when the RTD-LD is biased in region III, but now emitting downward–upward electrical pulses are observed, see figure 7.9(a), plot (v), and upward optical pulses that appear in bursts are obtained, see figure 7.9 plot (vi), due to the asymmetry of the I–V curve, which has a smaller slope on the valley side. Note that because the operating point is now at the low current value in region III, which defines the starting point of the excitable orbit, the system experiences a current increase, which explains the upward optical pulses. The analysis of the ISI, figure 7.9(b) (iii), reveals that the bursts manifest themselves in the histograms as a peaked structure at n/T_1, where T_1 = 350 ns corresponds to the excitable refractory time. This means that in this bursting regime, if we have an excitable orbit at T_1, then it is likely to be followed by another one at $T_2 = T_1 + T_1$. We also see the appearance of the statistics with its normal tail, which is very similar to figure 7.9(b) (i), as a result of the excitable response.

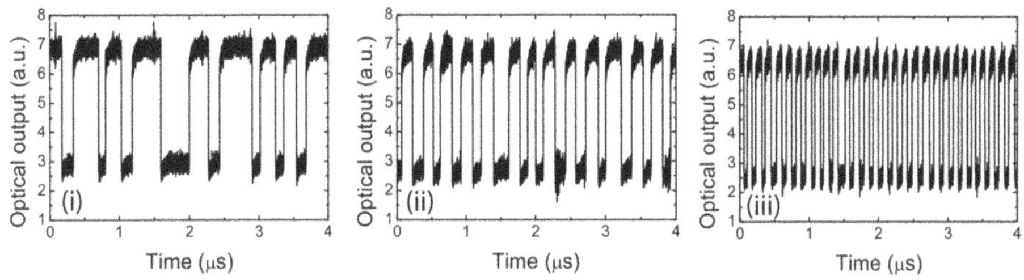

Figure 7.10. Experimental time traces of electrically noise induced neuron-like phenomena measured in the laser output for a circuit DC biased in a bistable condition, V_{dc} = 2 V. Random fired squared pulses at (i) V_{noise} = 125 mV, and (ii) V_{noise} = 150 mV; (iii) periodic sequence of square pulses, V_{noise} = 250 mV. Reprinted by permission from Springer Nature [25], Copyright (2014).

7.3.3 Bistable dynamics

When the RTD-LD is loaded with high resistance, see the circuit in figure 7.7(b), it shows a region of bistability around 2 V, see figure 7.7(d). This bistability is evidenced as noise-triggered hopping between the two stable states in regions I and III. A DC bias in such a condition results in switching between these two states, which leads to irregular waveforms composed of two plateaux whose duration is governed by the Kramer escape time in both basins of attraction, see figure 7.10, plots (i) and (ii). Increasing the noise level reduces the residence times in both states and leads to an almost regular square waveform, see figure 7.10(iii). This is the basis for stochastic resonance, which allows a low-amplitude signal to be amplified and regenerated in bistable systems by introducing just the right amount of noise.

7.4 Conclusions

In this chapter we have covered the nonlinear dynamics of semiconductor devices and systems based on resonant tunnelling, which included injection locking, period-adding, quasi-periodic route to chaos phenomena, as well as neuron-like spiking and bursting dynamics, and bistable dynamics. Fundamentally, resonant tunnelling results in a highly nonlinear I–V curve and a large bandwidth, and these attributes can be employed in a variety of ways to enable applications that include microwave-photonic systems for signal processing [1, 2], secure optical chaos communications [3, 4] and neuromorphic (brain-inspired) photonic applications [5]. Neuromorphic applications and tunnelling devices have a long association that extends back to 1962 and the work of Nagumo with the original Esaki diodes [22], which eventually evolved into the FitzHugh–Nagumo model of the neuron [23, 24].

The RTD and semiconductor device tunnelling phenomena generally continue to be a rich source of experimental examples of nonlinear dynamics concepts, which promise a large range of practical applications. In chapter 9, we will cover the excitable neuron-like properties of RTD photonic devices introduced here in the context of brain-inspired circuits for neuromorphic computing applications.

References

[1] Yao J P 2009 Microwave photonics *J. Lightwave Technol.* **27** 314–35
[2] Seeds A J 2002 Microwave photonics *IEEE Trans. Microwave Theory Tech.* **50** 877–87
[3] VanWiggeren G D and Roy R 1998 Communication with chaotic lasers *Science* **279** 1198–200
[4] Argyris A *et al* 2005 Chaos-based communications at high bit rates using commercial fibre-optic links *Nature* **438** 343–6
[5] de Lima T F, Shastri B J, Tait A N, Nahmias M A and Prucnal P R 2017 Progress in neuromorphic photonics *Nanophotonics* **6** 577–99
[6] Ott E 1993 *Chaos in Dynamical Systems* (Cambridge: Cambridge University Press)
[7] Strogatz S H 1994 *Nonlinear Dynamics and Chaos* (New York: Addison-Wesley)
[8] Adler R 1973 Study of locking phenomena in oscillators *Proc. IEEE* **61** 1380–5
[9] Romeira B *et al* 2009 Nonlinear dynamics of resonant tunneling optoelectronic circuits for wireless/optical interfaces *IEEE J. Quantum Electron.* **45** 1436–45
[10] Romeira B, Figueiredo J M L, Ironside C N and Slight T 2009 Chaotic dynamics in resonant tunneling optoelectronic voltage controlled oscillators *IEEE Photonics Technol. Lett.* **21** 1819–21
[11] Figueiredo J M L, Romeira B, Slight T J, Wang L, Wasige E and Ironside C N 2008 Self-oscillation and period adding from resonant tunnelling diode-laser diode circuit *Electron. Lett.* **44** 876–U207
[12] Yousefi M *et al* 2007 New role for nonlinear dynamics and chaos in integrated semiconductor laser technology *Phys. Rev. Lett.* **98** 044101
[13] Argyris A, Hamacher M, Chlouverakis K E, Bogris A and Syvridis D 2008 Photonic integrated device for chaos applications in communications *Phys. Rev. Lett.* **100** 194101
[14] Lindner B, Garcia-Ojalvo J, Neiman A and Schimansky-Geier L 2004 Effects of noise in excitable systems *Phys. Rep.-Rev. Sec. Phys. Lett.* **392** 321–424
[15] Hodgkin A L and Huxley A F 1952 A quantitative description of membrane current and its application to conduction end excitation in nerve *J. Physiol.* **117** 500–44
[16] Kuhnert L, Agladze K I and Krinsky V I 1989 Image-processing using light sensitive chemical waves *Nature* **337** 244–7
[17] Barbay S, Kuszelewicz R and Yacomotti A M 2011 Excitability in a semiconductor laser with saturable absorber *Opt. Lett.* **36** 4476–8
[18] Suzuki S, Shiraishi M, Shibayama H and Asada M 2013 High-power operation of terahertz oscillators with resonant tunneling diodes using impedance-matched antennas and array configuration *IEEE J. Sel. Top. Quantum Electron.* **19** 8500108
[19] Romeira B, Javaloyes J, Figueiredo J M L, Ironside C N, Cantu H I and Kelly A E 2013 Delayed feedback dynamics of lienard-type resonant tunneling-photo-detector optoelectronic oscillators *IEEE J. Quantum Electron.* **49** 31–42
[20] Hartmann F, Gammaitoni L, Hofling S, Forchel A and Worschech L 2011 Light-induced stochastic resonance in a nanoscale resonant-tunneling diode *Appl. Phys. Lett.* **98** 242109
[21] Romeira B, Javaloyes J, Ironside C N, Figueiredo J M L, Balle S and Piro O 2013 Excitability and optical pulse generation in semiconductor lasers driven by resonant tunneling diode photo-detectors *Opt. Express* **21** 20931–40
[22] Nagumo J, Arimoto S and Yoshizawa S 1962 An active pulse transmission line simulating nerve axon *Proc IRE* **50** 2061–70

[23] Izhikevic E M 2007 *Dynamical Systems in Neuroscience (Computational Neuroscience)* (Cambridge, MA: MIT Press)
[24] Izhikevich E M and FitzHugh R FitzHugh–Nagumo Model (http://www.scholarpedia.org/article/FitzHugh-Nagumo) (accessed October 2023)
[25] Romeira B *et al* 2014 Stochastic induced dynamics in neuromorphic optoelectronic oscillators *Opt. Quant. Electron* **46** 1391–6

Charlie Ironside, Bruno Romeira and José Figueiredo

Chapter 8

Resonant tunnelling diode-optoelectronic oscillator (RTD-OEO)

Chapter 8 covers an optoelectronic oscillator, the resonant tunnelling diode-optoelectronic oscillator (RTD-OEO), that uses some of the components discussed in early chapters, namely the resonant tunnelling diode-photodetector (RTD-PD) and the resonant tunnelling diode-laser diode (RTD-LD). These components, along with optical fibre delay lines, are employed to produce a high-Q oscillator with radio frequency (RF; ~1 GHz) and optical outputs. The RTD-OEO uses the optical fibre as part of a resonator to store and recirculate optical signals. This results in high-Q oscillators with exceptionally narrow linewidths. The RTD photonic components provide convenient on-chip electrical gain that can be used to compensate for optical loss in the resonator to achieve oscillation threshold. Here the RTD-OEO is described for the generation of high-purity microwave signals at multigigahertz frequencies and as a regenerative memory enabling optical buffering, storage, and reshaping of data signals.

8.1 Introduction

Over the last few decades, there has been intensive research on novel photonic-based oscillator technologies that are capable of generating microwave carriers in both the electrical and optical domains. Photonic-based oscillators have the potential to replace existing microwave resonator-based oscillator technologies, which cannot match all the advantages of photonic systems. There are numerous ways to generate a microwave signal in the optical domain, where it is possible to synchronise or control the oscillator for RF photonics applications using both electrical and optical signals. Figure 8.1 shows a few of the most studied methods based on either direct or external modulation of a semiconductor laser diode. The direct modulation example in figure 8.1(a) corresponds to one of the least expensive solutions. Figure 8.1(b) shows external modulation using a laser, an external electro-optic modulator, and a

doi:10.1088/978-0-7503-5714-2ch8

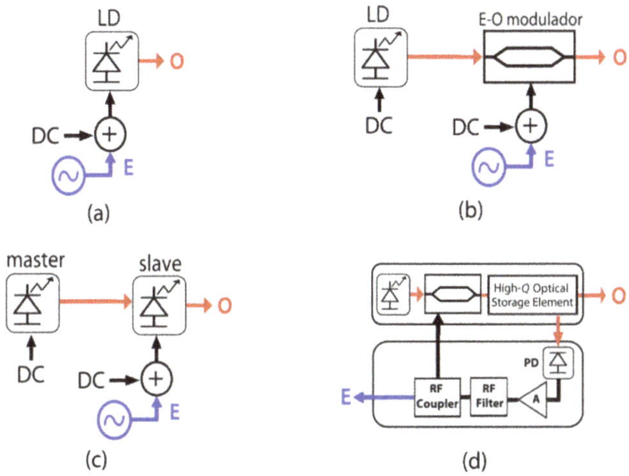

Figure 8.1. Methods for generating microwave electrical (E) signals in the optical (O) domain. (a) Direct modulation. (b) External modulation. (c) Optical injection locking of a directly modulated laser. (d) The optoelectronic oscillator (PD: photodetector; A: amplifier).

stable microwave signal from a local oscillator. This method enables higher performance while providing a trade-off between cost and energy consumption.

By using injection locking or delayed-feedback techniques, the generated microwave carriers from either direct or external modulation of a semiconductor laser can be significantly improved for higher performance. Some of the most popular methods are optical injection [1], optical feedback [2, 3] and optoelectronic feedback [4, 5]. In the case of the optical injection method, see figure 8.1(c), an external coherent light source in a master–slave configuration is used to provide bandwidth enhancement, chirp and noise reduction for optical communication applications [6]. In the optical feedback method (not shown), an external mirror is used to optically reinject a fraction of the light emitted by the laser back into its active region. The optical delay in the system is introduced by the round-trip time of light propagation in the external cavity. In the optoelectronic feedback method, a fast-responding photodiode detects the light output, followed by current reinjection back to the laser's RF input. This generates a modulated current signal that is an image of the laser's output and is added to the laser's injection current after an optoelectronic delay time.

The optoelectronic oscillator, shown in figure 8.1(d), is a unique example of the implementation of the optoelectronic feedback method. In this case, the laser diode provides a continuous wave light source, while the nonlinear elements are external electronic and/or electro-optic components (e.g. electro-optic modulators, RF oscillators, etc). Several examples of time-delayed OEO topologies combining electronic and photonic components have been extensively reported over the last two decades, showing the most promising results for the generation of stable and high-purity RF signals, which will be the focus of this chapter.

Following a general introduction to OEO systems, this chapter describes the experimental setup of a fibre-based optoelectronic oscillator architecture based on self-synchronised resonant tunnelling diode photonic integrated devices, followed by the key results that were recently obtained operating this architecture in either single- or dual-fibre loop configurations. The experimental results are compared with numerical simulations of a Liénard oscillator model with time-delayed feedback. Finally, applications utilising time-delayed-feedback RTD-OEOs are discussed, including clock generation and optical data buffering. This chapter is an adaptation of chapter 5 of [7], which has been thoroughly revised and updated.

8.2 Optoelectronic oscillator

An optoelectronic oscillator (OEO) is a hybrid electro-optical feedback oscillator that converts optical energy from a laser to microwave signals. Instead of a resonant circuit, the OEO stores energy in the form of a long, low-loss fibre-optic delay line. The first highly-stable OEO system was demonstrated in 1996 at the Jet Propulsion Laboratory (for a perspective review, see [8]). It was made of a multi-kilometre-long optical fibre delay line, a wide-band electro-optical modulator, light from a semiconductor laser coherent light beam, a photodiode, a narrow microwave RF filter, and microwave and/or optical amplifiers. Its basic architecture is shown in figure 8.1(d). This is one of the most studied configurations. The conditions for self-sustained oscillations are coherent addition of partial waves each way around the loop (the loop phase shift must be an integer multiple of 2π) and a loop gain larger than the losses for the circulating waves in the loop. This forms the Barkhausen criteria [9] required to achieve stable oscillations in the OEO. Because of the fibre's extremely low losses and large energy, a high loaded resonator quality factor can be obtained, resulting in oscillations with extremely low phase noise. Therefore, OEOs have fascinating applications in optical communication links, as well as in precise test and measurement equipment.

Several alternative OEO topologies to the one presented in figure 8.1(d) have been reported in recent years, showing stable and low phase-noise RF carriers using either direct or external modulation of laser diodes [10], or heterojunction bipolar transistors [11], to name a few. Nevertheless, to obtain RF signals in both the electrical and optical domains, additional electronic and optoelectronic components are required, such as phase-locked local oscillators, amplifiers, photo-detectors, light emitters and/or amplifiers or modulators. In addition, techniques such as bidirectional injection or self-injection locking may be required to achieve the extremely low phase-noise levels that are required by some applications. As a result, these OEO alternatives are rather complex and costly for applications such as radio-over-fibre networks, where compact solutions for RF signal transmission are required [12, 13].

Although the laser source can influence OEO performance (specifically, the impact of laser intensity noise on the phase noise), most conventional schemes require the use of an external modulator (typically a Mach–Zehnder) to introduce a microwave sub-carrier into the optical signal. This places a practical maximum frequency of oscillation, limited by the bandwidth of the electro-optic modulator, although state-of-the-art modulators are able to operate at ultrafast speeds. Importantly, to obtain a highly stable and single oscillation signal, a filter in the

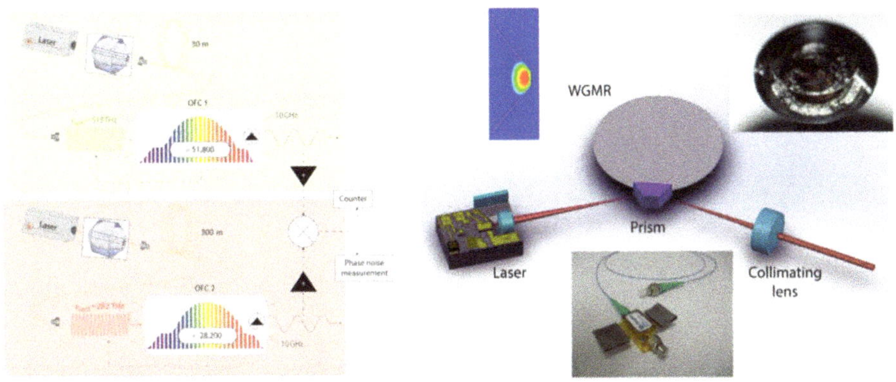

(a) (b)

Figure 8.2. State of the art implementations of ultralow phase noise RF signals. (a) Schematic of the experimental setup used to generate and characterise the 10 GHz low-noise microwaves via optical frequency division [14]. (b) Miniature external cavity semiconductor laser for ultralow noise signal generation [15]. Light from the pump laser semiconductor DFB laser enters the WGM resonator through the prism. Part of light is reflected back to the laser because of Rayleigh scattering in the resonator. The light that exits the prism is collimated and used for processing. The insets show power distribution in the resonator mode (top left-hand side), a picture of the actual resonator (top right-hand side) as well as a picture of the packaged laser (bottom). Reprinted by permission from Springer Nature [14], Copyright (2011) & Reproduced from [15]. CC BY 4.0.

feedback loop is required to remove the harmonic signals generated by the nonlinear response of the electro-optic modulator.

To address the limitations imposed by OEOs, namely mode selection filtering, various approaches to selecting a single operational mode from those available have been proposed. The most promising implementations employ photonic generators via optical frequency division [14], as shown in figure 8.2(a), and laser self-injection locking via optical feedback from high-Q whispering gallery mode (WGM) resonators [15], as shown in figure 8.2(b). These methods provide ultra-stable electrical signals comparable to those produced by the best microwave oscillators and clearly demonstrate the benefits of using optical methods to generate high-frequency and low phase-noise microwave carriers. The single-sideband (SSB) phase noise for several leading microwave generation technologies in the 10 GHz range is summarised in figure 8.3.

Although remarkable results have been achieved, almost all of the aforementioned methods are complex, costly, and require a large number of electronic and optical high-bandwidth amplification components that consume large amounts of energy. In fact, amplification processes are also significant noise contributor elements. Thus, a great deal of work on OEOs has been devoted to providing high-performance and high-power microwave and millimetre-wavelength oscillators that are not only miniature in size but also consume far less power than existing devices.

In what follows, we will discuss our recent work on a fibre-based optoelectronic oscillator architecture consisting of self-synchronised resonant tunnelling diode

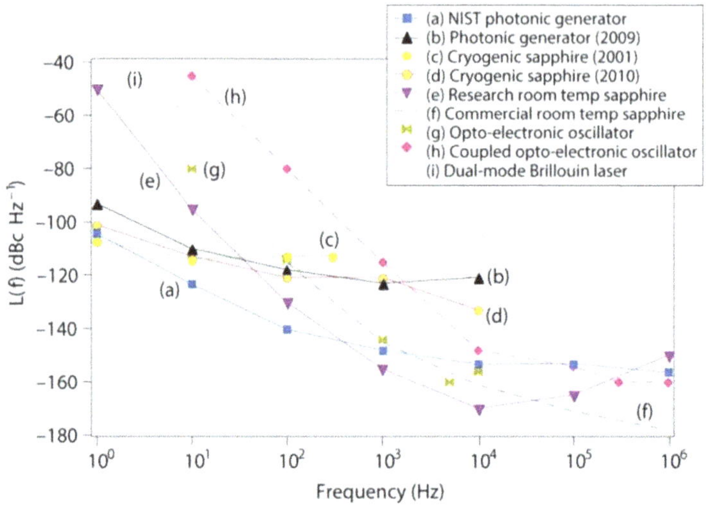

Figure 8.3. SSB phase noise for several leading microwave generation technologies in the 10 GHz range [14]. (a) Result of the state-of-the-art NIST photonic generator. (b) Previous Er:fibre and Ti:sapphire optical frequency divider results. (c and d) Cryogenic sapphire oscillators. (e) Research room-temperature sapphire oscillator (f), commercial room-temperature sapphire oscillator. (g and h) Long fibre (g) and coupled (h) optoelectronic oscillators. (i) Upper limit of a dual-mode Brillouin laser. Reprinted by permission from Springer Nature [14], Copyright (2011).

photonic integrated devices, which offers a new and alternative solution for compact photonic integrated circuit-based OEOs.

8.3 Photonic integrated RTD-OEO

RTD-based oscillators typically have high noise levels, with single-sideband noise power densities significantly larger than -100 dBc Hz^{-1} at 100 kHz offset from the carrier frequency. This is in comparison to the typical noise levels of other analogous electronic free-running oscillators, which were recently presented in [11]. External injection locking techniques, as discussed in chapter 7 of this book, are commonly used to achieve further improvements in the phase noise. The RTD-PD-based OEO topology discussed here is a reliable method for ensuring the stability of free-running oscillations, while reducing fluctuations and noise without the use of bulky external reference sources.

In this section, we discuss the feasibility of using RTD-PD oscillators for low powered photonic integrated OEO systems. This approach simplifies the reported OEO configurations reviewed in the previous section and shown in figure 8.4(a). The architecture of a photonic integrated RTD-OEO is depicted in figure 8.4(b). It consists of an RTD-PD oscillator that operates at multi-gigahertz frequencies. The microwave signals that are produced modulate directly the current of a laser diode. A low-loss high-Q cavity provided by the optical fibre delay line in a feedback loop configuration enables high-purity microwave signals. The RTD-OEO does not

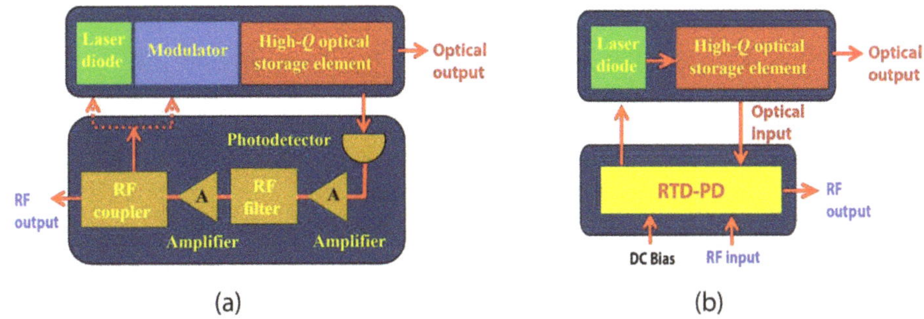

Figure 8.4. Optoelectronic oscillator (OEO) configurations. (a) First OEO architecture of an optoelectronic oscillator. (b) A self-synchronised OEO using a low-loss high-Q cavity and an RTD-PD to drive a laser diode on time-delayed-feedback configuration.

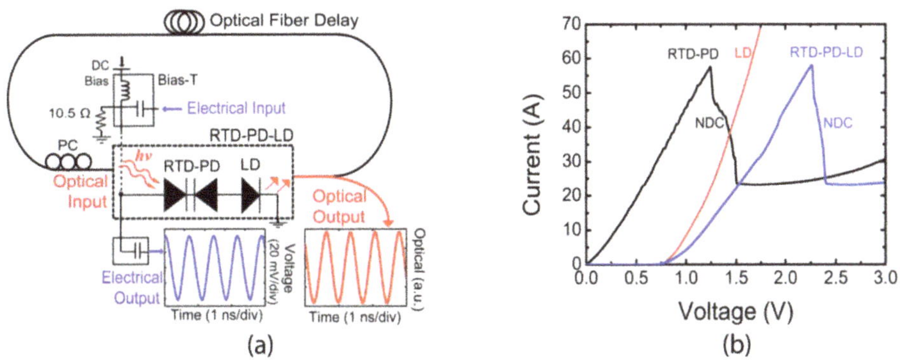

Figure 8.5. Self-synchronization circuit system. (a) Schematic setup of the self-synchronised RTD-OEO. (b) I–V characteristics of the RTD-PD-LD, RTD-PD and laser diode devices, showing the respective negative differential conductance regions.

require low-noise and stable external RF amplifiers nor microwave filters, which are typically required to compensate for E/O and O/E conversion losses and frequency mode selection, respectively. By removing the RF and erbium doped fibre amplifier (EDFA) amplifiers from the OEO system, the main sources of noise in an OEO are removed. Furthermore, the RTD-OEO requires less power than previous OEOs and does not require additional high-speed, high-cost nor bulky components such as photo-detectors or modulators. The following sections present two main architectures: (i) the single-optical fibre loop, and (ii) the dual-optical fibre loop configuration. The stability and phase-noise characteristics as a function of optical power and delay line length are discussed.

8.3.1 Single-optical fibre loop

Figure 8.5(a) shows a closed-loop RTD-OEO implementation consisting of a delay line implemented using a single-mode optical fibre roll, a DBQW-RTD monolithic waveguide photodetector oscillator and a commercial laser diode source. We obtain an RTD-OEO with both electrical and optical input and output ports using this simple feedback route configuration. The RTD-OEO frequency and phase-noise fluctuations are significantly reduced after passing through the high-Q optical delay line. The use of fibre-rolls of different lengths provides an easy way of varying the optical feedback route of the RTD-OEO, but other more compact delay lines, such as WGM resonators, could be considered.

In the RTD-OEO under consideration here, both the RTD-PD and the laser diode are hybrid integrated on a printed circuit board transmission line layout, as described in Chapters 6 and 7. The RTD-PD devices consist of a chip die measuring approximately $0.5 \times 0.4\ \text{mm}^2$ (see chapter 5 for details). The RTD photodetector's typical optical responsivity was $\sim$0.28 A W^{-1} at 1.55 μm. These values were measured while the RTD-PD was DC biased around 2.4 V, i.e., when operating near the valley region, as shown in figure 8.5(b). The microwave signals are converted electro-optically via direct modulation of a laser diode operating around 1.55 μm with an efficiency of $\sim$0.25 W A^{-1}. This arrangement, as shown in figure 8.5(a), provides a simple configuration that does not require external signal amplification. However, because the RTD-PD devices were not optimised for this particular experiment, EDFA amplifiers were used in some experiments, specifically the dual-loop fibre experiments presented in the following subsection, to compensate for the losses of coupling the light in more than one optical fibre.

The RTD-OEO operates as a self-synchronised optoelectronic voltage-controlled oscillator, which means that the free-running frequency of the RTD-OEO is controlled by adjusting the DC bias voltage. When biased in the negative differential conductance (NDC) region, electrical free-running current oscillations are formed, directly modulating the laser diode. The LD optical modulated output is then coupled to a lensed single-mode fibre, routed through an optical fibre delay line, and finally reinjected into RTD-PD waveguide optical input via an identical single-mode lensed fibre. The optical reinjection mechanism enables self-injection locking of the RTD free-running oscillations. The RTD-based OEO's frequency is determined by the RTD-PD parallel capacitance and the series equivalent inductance of the gold electrical wires used to connect the RTD-PD to the laser diode, with typical wire lengths of around 4 mm, allowing us to estimate an equivalent inductance of 3.6 nH. The photonic integrated circuit discussed here can operate at frequencies of up to 1.4 GHz. The oscillator could operate at >10 GHz using an optimised circuit configuration for layouts that significantly reduce the series inductance. The laser is the device that limits the maximum frequency bandwidth that can be achieved with this OEO configuration because it has a 3-dB cut-off frequency of 11 GHz. To achieve frequencies well above the maximum bandwidth available in laser components (>30 GHz), an RTD-EAM device, discussed in chapter 4, or some other external modulator device can be used to modulate the light intensity.

The RTD-OEO circuit discussed here oscillates with a free-running frequency that is tuneable from 1.05 to 1.41 GHz (by tuning the DC bias) in the configuration shown in figure 8.5(a), without optical reinjection. The frequency of the electrical and optical modulated free-running oscillations was 1.4052 GHz, the oscillations in

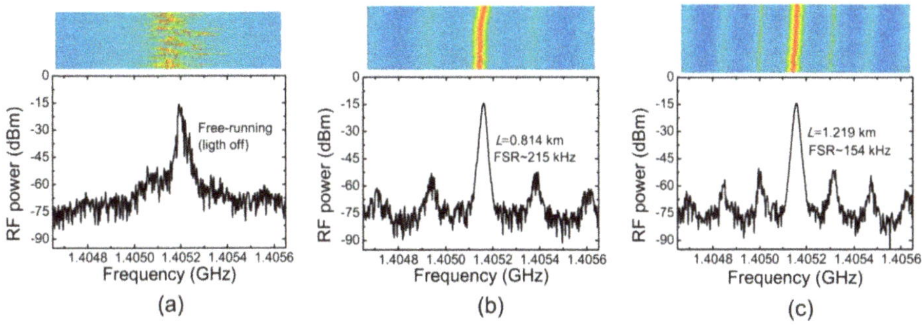

Figure 8.6. RTD-OEO output showing RF power spectrum traces. (a) Free-running. (b–c) Self-synchronisation at $P \sim 5$ dBm using fibre loops of (b) $L_f = 0.814$ km, and (c) $L_f = 1.219$ km. All figures had their frequency span and resolution bandwidth set to 1 MHz and 10 kHz, respectively. The top portion of each figure shows a spectrogram representing the evolution of spectral density recorded over 30 s (time in the vertical axis).

time domain are shown in the inset of figure 8.5(a), when biased close to the valley region, ~2.4 V, with an estimated power consumption of 0.55 W. The power consumption can also be further reduced to less than 100 mW by connecting a capacitor in series with the shunt resistor shown in figure 8.5(a). Figure 8.6(a) depicts the free-running oscillation spectrum trace, which demonstrates the typical broad spectrum caused by the noise-induced fluctuations of the fundamental frequency oscillation. The RTD circuit, the *I*–*V* characteristic, and the electrical losses of the measurement setup all contribute to a reduction of the overall measured electrical power of around −15 dBm.

The performance of the RTD-OEO was investigated using fibre delay lines with lengths, L_f, ranging from 0.4 to 4.4 km. The spectra of the self-synchronised free-running oscillations using fibre-roll lengths of 0.8 and 1.2 km are shown in figure 8.6(b) and (c), respectively. The average in-fibre optical power was P~5 dBm. The spectrum surveillance, which shows the plot density spectrum, is also presented inset on top of each figure. The results show that the self-injection improves signal quality and frequency stability while significantly reducing phase noise.

Figures 8.6(b) and (c) show side modes separated from the centre frequency by about 215 and 154 kHz, respectively, with single mode suppression ratios (SMSRs) of 38 and 36 dB, respectively. These correspond to the mode spacing of the free spectral range (FSR) of the RTD-OEO and are related to the time delay, τ_d, which in this case is determined by the electrical time delay, τ_e, and the optical time delay, τ_{op}, which is set by the length of the optical fibre. The optical time delay τ_{op} is much larger than the electrical time delay τ_e in the experiments discussed here. As a result, the FSR of the RTD-OEO can be calculated as follows:

$$\text{FSR} = \frac{1}{\tau_e + \tau_{\text{op}}} \cong \frac{1}{\tau_{\text{op}}} \cong \frac{c}{n_F L_f} \tag{8.1}$$

where $\tau_{\text{op}} = n_F L_f / c$, n_F is the effective refractive index of the optical fibre and c is the velocity of light, implying that the RTD-OEO mode spacing is determined by the fibre length. The following FSR values are estimated from equation (8.1): $L_f = 0.814$

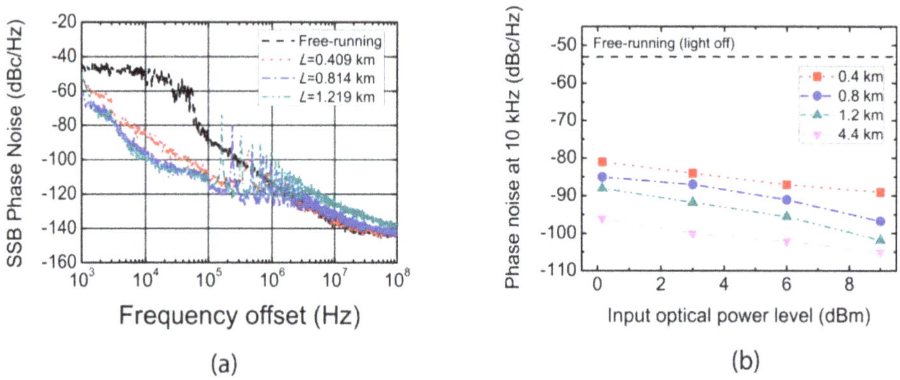

Figure 8.7. Injection locking experimental results. (a) RTD-OEO SSB phase-noise measurements for the free-running oscillation case without feedback loop (dashed black line), and for the closed-loop case using various fibre-roll lengths and a self-injected optical power of $P \sim 9$ dBm. (b) RTD-OEO SSB phase-noise measurements as a function of in-fibre optical power at 10 kHz offset from the centre frequency (the free-running value is shown as the dashed black line).

km, FSR = 251 kHz; and $L_f = 1.219$ km, FSR = 167.6 kHz, which compare to the experimental values for lengths shown in figures 8.6(b) and (c), respectively.

The respective SSB phase noise of the output signals were measured to verify the phase-noise reduction performance, as shown in figure 8.7. The corresponding self-injection locked phase-noise values for an in-fibre optical power of P~9 dBm and optical fibre lengths of 0.409, 0.814 and 1.219 km, are −89, −96.7 and −101.8 dBc Hz^{-1}, respectively, as shown in figure 8.7(a). When compared to the free-running oscillations without light reinjection, a phase-noise reduction of more than 40 dB at 10 kHz offset is achieved. As shown in figure 8.7(b), the SSB phase noise at 10 kHz offset from the centre frequency is substantially reduced in the self-synchronised mode as the in-fibre optical injection power and fibre delay are increased.

The results of SSB phase noise for a very long delay line, i.e., using a long fibre roll with $L_f = 4.4$ km, are also shown in figure 8.7(b). At frequencies offsets below the side modes, a phase-noise reduction of up to 50 dB is achieved when compared to free-running oscillations without light reinjection. At power levels close to 3 dBm, phase-noise values less than −100 dBc Hz^{-1} at 10 kHz offset are achieved, representing a ~17 dB reduction over $L_f = 0.4$ km. Furthermore, as the reinjected power level increases, the optical loop losses are reduced. However, the side modes become more visible. As discussed next, the introduction of a multi-loop fibre configuration provides an excellent method for mode selection because it enables low phase noise and frequency stability due to the very long delay line, while suppressing unwanted signals that deteriorate the performance of the RTD-OEO around frequency offsets close to the FSR of the delay line.

8.3.2 Dual-optical fibre loop

As demonstrated in the preceding subsection, a disadvantage of using a single-fibre loop is the appearance of side mode oscillations. These modes, which are highly undesirable for some applications, are caused by wave propagation multiple times around the OEO loop. There are two effective methods for removing or suppressing side modes. One approach involves using multiple loops of fibre to create a narrow band filter. A high-*Q* optical cavity can also be used to filter out unwanted modes.

Yao and Maleki demonstrated a dual-loop OEO configuration in which the optical power was split by an optical decoupler and the optical signal was detected by two photo-detectors [16]. After amplification, the two converted RF signals were sent to an RF coupler and fed to the modulator. The process was equivalent to a single-loop OEO in each loop. In this configuration, the method takes advantage of the natural structure of the OEO cavity to obtain filtering via an additional fibre loop or loops of shorter lengths connected in parallel with the longer fibre loop. Because of the gain competition between the two sets of modes, the side modes are efficiently suppressed when the two loops work together. In this case, because the RF signals have the same oscillation frequency and are coupled in the electrical domain, there is no random interference or beat noise. Although this design gave the desired results, two high-speed photo-detectors were still needed.

In our configuration, significant improvements in side mode reduction can also be achieved. In this subsection, we discuss a carrier suppression scheme in a dual-loop RTD-OEO. In this setup, the laser optical output is split into two parts that propagate through two fibers with different lenghts before being combined in a single fibre that is coupled to a single RTD-PD. This means that the RF signals are coupled in the optical domain without the use of additional active electrical devices. Because of the interference between the two combined optical signals, the dual-loop RTD-OEO enables high mode selectivity while suppressing side modes. The RTD-OEO circuit configuration and components used in the dual-loop experiment were identical to those presented in the single-fibre experiment. The circuit under analysis here generated oscillations with natural frequencies ranging from 0.944 to 1.129 GHz, depending on direct current (DC) bias. In this experiment, the RTD-OEO was DC biased at 2.33 V (close to the valley), producing oscillations at around 1.121 GHz. When biased in the valley region, the RTD-PD device had typical optical responsivity of 0.25 A W^{-1} at 1550 nm.

The effect of the dual-loop RTD-OEO was studied experimentally by measuring the phase noise and side modes characteristics using fibre loop lengths of 4.4 and 0.4 km, see figure 8.8. The longer fibre enables high spectral purity and low phase noise, whereas the shorter fibre suppresses the side modes near the carrier frequency. When either path of the fibre is broken, there is a complete single-loop OEO that can run freely, as similarly demonstrated in the previous subsection. The mode spacing was 425 kHz with the long path broken (4.4 km fibre loop), and the measured side mode suppression ratio was −53 dBc at $P{\sim}6$ dBm, see figure 8.8(a). The mode spacing was 45 kHz with the short path broken (0.4 km fibre loop) and the measured side mode suppression ratio was around −20 dBc, see figure 8.8(b). The dual-loop configuration improved the side mode suppression ratio to −60 dBc, see figure 8.8(c). Side modes can be effectively suppressed by 20–40 dB when compared with the single loop results, depending on the input optical power launched into the RTD-PD.

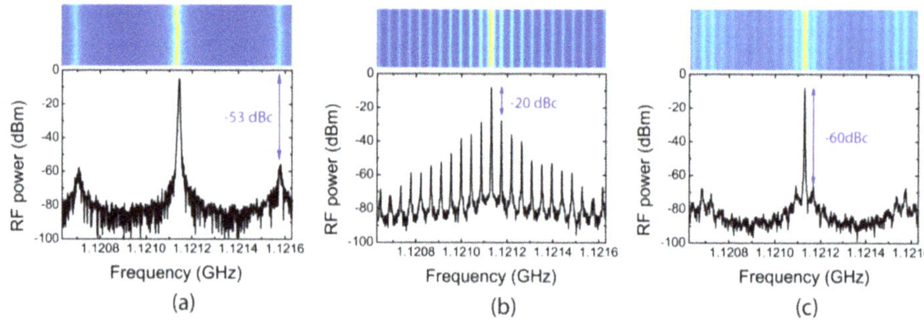

Figure 8.8. RF power spectra output of the RTD-OEO showing the SMSR for (a) 0.4 km, (b) 4.4 km single loop and (c) dual loop configurations. The top part of each figure shows a spectrogram representing the evolution of spectral density recorded over 1 min (time in the vertical axis).

The SSB phase noises of the single- and dual-loop configurations were measured. The phase noise of the dual-loop RTD-OEO, measured at 10 kHz away from the carrier, was -99.28 dBc Hz^{-1} at 6 dBm in-fibre reinjected optical power, and -102.88 dBc Hz^{-1} at 9 dBm in-fibre reinjected optical power, which compares to -101.67 dBc Hz^{-1} for the single loop at 4.4 km. Therefore, in addition to maintaining the quality of the phase noise below -100 dBc Hz^{-1} at lower frequency offsets, successful suppression of the side modes was achieved in the spectral region above 45 kHz frequency offset of the carrier frequency, confirming the efficient side mode suppression capability of the dual-loop configuration. The measurements also revealed that the dual-loop configuration introduces additional close-to-carrier noise, resulting in slightly higher noise levels at lower carrier offsets when compared to the single 4.4 km long path.

As a final note, the dual-loop OEO implementation discussed here uses only a single RTD-PD, which significantly reduces the number of components typically required to form a dual-loop OEO configuration. However, it may encounter issues related to optical coupling, such as interference and beating because the RF signals are coupled in the optical domain while in configurations presented elsewhere the coupling is accomplished electrically. An effective and simple way to avoid these two effects while retaining the dual-loop function is to use, for example, a polarisation beam splitter and combiner. The fluctuation of polarisation and loop length fluctuations are additional challenges of implementing the dual-loop RTD-OEO scheme. For example, when the polarisation deviates significantly, the optimised power distribution among the fibres is destroyed and the side mode suppression is less effective. Polarisation-maintaining fibres can be used in the system to eliminate this challenge. In the case of the intensity fluctuations caused typically by environmental changes, these effects are difficult to overcome unless temperature control or feedback control is employed.

In summary, the results of both single- and dual-loop schemes demonstrate that self-synchronisation substantially improves the phase quality of free-running output signals, with a compromise between the length of the time-delayed-feedback loop and the phase-noise suppression. Although the phase-noise characteristics of

RTD-OEOs are not comparable to commercially available voltage-controlled modules, such as coaxial resonator oscillators (typically with phase noise of −120 dBc Hz^{-1} at 10 kHz offset for 1 GHz operation), it is expected that further techniques such as temperature and vibration control will have a significant impact in the RTD-OEO performance. Notably, the results presented here were limited by the responsivity of non-optimised RTD-PDs, which can be improved further by using anti-reflection coating facets and maximising waveguide coupling efficiency and RTD detector quantum efficiency. Finally, since there is still significant scope for improvement in many aspects of the OEO configuration, such as monolithic integration of the RTD-PD with laser diode with an on-chip high-Q feedback route, OEO topologies based on self-injection locking of RTD oscillators may be potential alternatives for novel low-cost and on-chip OEOs with a small footprint and significantly lower energy consumptions than existing devices.

8.3.3 Time-delayed-feedback Liénard oscillator model

In chapters 6 and 7, where the theory of RTD and laser diode dynamical systems was discussed, we considered systems that can be analysed by a set of first-order ordinary differential equations. The RTD-OEO system belongs to a different class of systems, i.e., delay systems. Time delayed-feedback systems, where the dynamic characteristics of a system at a given instant t depend on the state of the system at an anterior time $t - \tau_d$, where τ_d is the time delay, are common in scientific fields such as physiology, biology, epidemiology, and control systems [17]. The effect of the delayed feedback is to reinject one or more state variables (e.g. current or intensity) into the system, with a delay. Depending on the time delay and feedback strength, a delayed-feedback loop can be designed to (i) achieve highly-stable periodic oscillations, or (ii) exceedingly complex dynamical phenomena and aperiodic oscillations, namely chaos [18, 19]. Notably, the dynamical characteristics of such systems can be easily synchronised and controlled via external electro-optical perturbations [20, 21]. The time-delayed-feedback RTD-OEO system is modelled by a system of first-order delay-differential equations:

$$\frac{d\boldsymbol{x}(t)}{dt} = \boldsymbol{F}(\boldsymbol{x}(t), \boldsymbol{x}(t + \tau_d); \boldsymbol{p}) \tag{8.2}$$

where τ_d is the delay of the system and $\boldsymbol{F}$ is a vector field. The derivative of $\mathbf{x}(t)$ at time t depends on the value of $x(t)$ at time t and on the value that $x(t)$ took in the past at time $t+\tau_d$. This means that it is necessary to know the vectorial function on the time interval of length τ_d to determine the state of the system and its evolution because the phase space is an infinite-dimensional functional space. The set of parameters $\boldsymbol{p} = (p_1; p_2; \ldots)$ describe the features of a particular system modelled by the vector field. When the vector field depends explicitly on time, the system is said to be non-autonomous.

In this section, we discuss a numerical model of the single-loop delayed-feedback RTD-OEO system using delay differential equations (DDEs) [22]. This model includes a time-delayed-feedback Liénard oscillator that models the RTD electrical

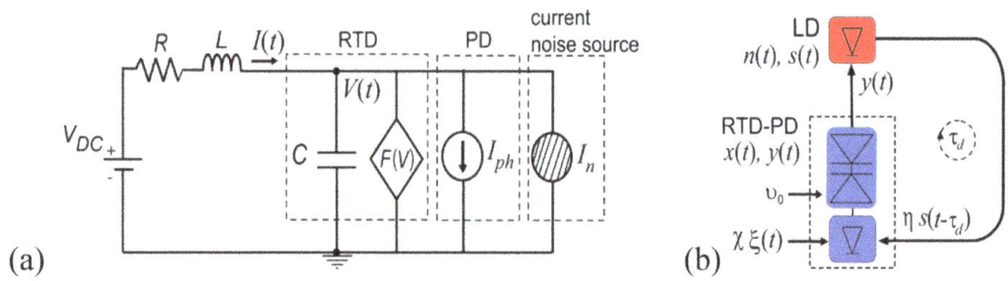

Figure 8.9. Time-delayed feedback circuit system. (a) Equivalent electrical circuit schematic of the resonant tunnelling diode photodetector oscillator with current noise injection. The resistor R and inductor L account to the circuit series and parasitic resistance and inductance of RTD-PD-LD circuit, respectively. (b) Block diagram of the mathematical representation of the delayed-feedback Liénard oscillator-laser diode dimensionless system.

dynamics and the feedback loop of the OEO, laser rate equations describing the laser diode dynamics and white Gaussian input noise to describe the dominant OEO noise sources, i.e., the thermal and shot noise.

The dynamics of the RTD-OEO are analysed considering the lumped electrical circuit of figure 8.9. This circuit is equivalent to a Liénard oscillator [23, 24], which includes photocurrent (I_{ph}) and current noise (I_{n}) terms. The RTD-PD is represented by its intrinsic capacitance in parallel with a voltage dependent current source $F(V)$, and its photo-detection characteristics are modelled as an optical power dependent current source, I_{ph} [7]. (See chapter 5 for more details on the photodetector model.)

Using Kirchhoff's rules, the same procedure discussed in chapters 6 and 7 was applied to obtain the two dimensionless first-order differential equations that describe the electrical circuit of figure 8.9(a). Then, the electrical system was coupled to the laser diode rate equations that model the laser system. As previously, time is normalised to the characteristic frequency of oscillation of the RTD-PD-LD, $\omega_0 = (\sqrt{LC})^{-1}$, hence $t = (\omega_0)^{-1}\tau'$. Finally, by redefining τ' as t and introducing into the system the optical delayed feedback $s(t - \tau'_d)$, note that η is the feedback strength and τ'_d is the time delay with respect to the dimensionless time t, the Liénard oscillator-laser diode dynamical system is transformed into the following dimensionless coupled DDEs:

$$\frac{dx(t)}{dt} = \frac{1}{\mu}[y(t) - f(x) - \chi\xi(t) - \eta s(t - \tau'_d)] \tag{8.3}$$

$$\frac{dy(t)}{dt} = \mu[\nu_0 + \gamma y(t) - x(t)] \tag{8.4}$$

$$\frac{dn(t)}{dt} = \frac{1}{\tau'_n}\left[\frac{i_m(t)}{i_{th}} - n(t) - \frac{n(t) - \delta}{1 - \delta}\{1 - \epsilon s(t)\}s(t)\right] \tag{8.5}$$

$$\frac{ds(t)}{dt} = \frac{1}{\tau'_p}\left[\frac{n(t) - \delta}{1 - \delta}\{1 - \epsilon s(t)\}s(t) - s(t) + \beta n(t)\right] \tag{8.6}$$

These equations represent the optical feedback control of the RTD-PD-LD through the variable $s(t - \tau'_d)$. The feedback strength η parameter depends on the fraction of the laser optical output power introduced into the delayed-feedback loop and on the RTD-PD detection characteristics. The function $f(x)$ comes from the normalisation of $F(V)$, and $\mu = V_0/I_0\sqrt{C/L}$ is dimensionless. Equations (8.5) and (8.6) are the dimensionless rate equations describing laser diode normalised photon $s(t)$ and injected carrier $n(t)$ densities, respectively, where $\delta = \frac{N_0}{N_{\rm th}}$ and $\epsilon = \epsilon_n S_0$ are two dimensionless parameters (where N_0 is the carrier density for transparency, $N_{\rm th}$ is the carrier density at threshold, ϵ_n is the value for the nonlinear gain compression factor and $S_0 = \Gamma(\tau_p/\tau_n)N_{\rm th}$ where Γ is τ_ps the photon lifetime and τ_n is the carrier lifetime $i_m(t)$ accounts for the bias current and the current oscillations produced by RTD-PD electrical model, Equations (8.3)–(8.4), and $i_{\rm th}$ is the dimensionless laser diode threshold current. The parameters τ_p' and τ_n' come from the time rescaling. Figure 8.9(b) presents a block diagram of the mathematical representation of the delayed-feedback system and respective variables and key parameters.

In real systems, the unavoidable noise sources affect the dynamics by introducing amplitude and period fluctuations, even in the most stable periodic signals. In our case, several sources or random processes are at play, e.g., thermal and shot noises. We model their overall effect as an effective delta-correlated Gaussian white noise of zero mean input source $\chi\xi(t)$, equation (8.3), where the parameter χ is the dimensionless variance of the distribution and denote the noise strength.

In what follows, we compare the delayed-feedback Liénard OEO model in the single loop configuration and compare with experimental results. When the time-delayed feedback is included, important dynamical effects are observed that include close-to-carrier noise reduction and the appearance of side modes due to the delay contribution, similar to the experimental results reported previously in the single loop configuration.

Figure 8.10 presents the experimental (a) and simulated (b) power spectra around the fundamental free-running oscillation with and without delayed feedback. Figure 8.10(a) presents the experimental results of self-synchronised electrical output using an in-fibre optical reinjected power P of ~ 6 dBm and 0.4 km fibre length. The results show phase noise and linewidth reductions at offsets below 250 kHz of the carrier frequency, and side modes separated by about 425 kHz with a SMSR of −43 dBc. Note that the linewidth of the self-synchronised output could not be resolved by the 3 kHz resolution limit set by the instrument. Figure 8.10(b) shows the corresponding simulation for a time delay of $\tau_{\rm d} = 2.35$ μs and a feedback strength to noise ratio of $\theta = 2.0$ (corresponding to a feedback strength $\eta = 1 \times 10^{-3}$, for an injected optical power of 6 dBm). As observed experimentally, the introduction of a time-delayed feedback variable in the Liénard OEO model narrows the linewidth of the fundamental oscillation frequency and generates frequency side peaks due to the time delay, which corresponds to the FSR of the RTD-OEO system. The SMSR of −44 dBc in the simulation compares with the experimental result shown in figure 8.10(a).

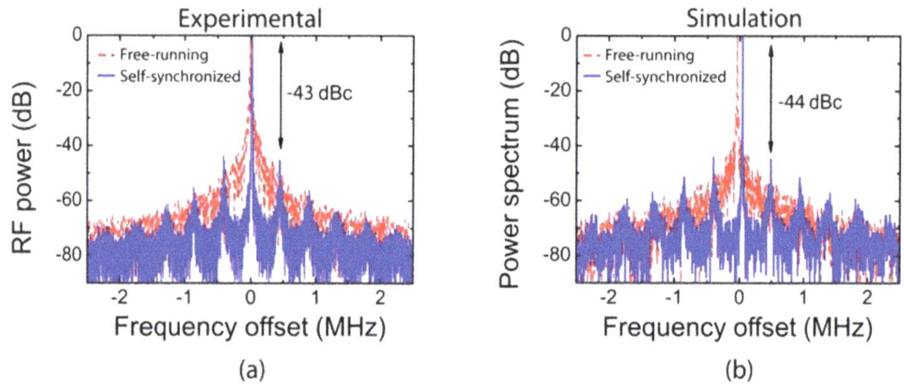

Figure 8.10. Self-synchronization results. (a) Experimental RF power spectra of free-running oscillation and self-synchronised electrical output at $P \sim 6$ dBm and 0.4 km fibre length. (b) Simulated power spectra of free-running fundamental oscillation and self-synchronised $x(t)$ outputs with time delay of $\tau_d = 2.35$ μs and feedback strength to noise ratio of $\theta = 2.0$. In both plots the span was 5 MHz and the central frequency was 1.12207 GHz.

The dynamical properties of the OEO model investigated here follow the dynamics observed experimentally where the characteristics of the OEO can be controlled using optical delay lines. The large number of side-bands spaced by FSR (inversely proportional to the time delay) is an indication that more complex dynamics may occur, which is strongly depended on the feedback level and the length of the external cavity. In the results presented here, considering the moderate levels of feedback employed, $\eta \ll \Delta I/I_0$, where ΔI is the peak-to-valley current ratio, only stable self-synchronised oscillations were observed. This model can be used either to map the dynamical regimes of RTD-OEOs in order to avoid unstable signals or to explore the rich dynamical characteristics, which can be exploited for innovative applications such as generation of frequency combs [5], or even chaotic transitions [25] controlled by the delayed-feedback parameter.

8.4 Regenerative RTD-OEO

Novel methods of buffering optical signals are highly desirable to avoid congestion of information traffic and to realise efficient optical interconnects. Recently, all-optical buffer memories have been proposed based on slow-light optical delay lines or using the Kerr nonlinearity on a standard silica optical fibre providing functionalities such as all-optical storage and reshaping. However, for realistic applications, an optical buffer has to be compact for on-chip integration, which rules out most existing schemes because they are not easily scalable to a millimetre size footprint, and the writing process is often complex and costly. Therefore, configurations that combine the robustness of semiconductor nanoelectronics and optical devices with the wide-bandwidth of photonic elements offer significant advantages because they can provide small size, high-speed and low-cost alternatives to the all-optical buffers currently proposed.

In this section, we discuss the implementation of regenerative photonic memories using the RTD-OEO system described in previous sections. However, in this case the optical delay line with a time delay inserted off-chip provides a mechanism for the reinjection of fired excitable pulses, a type of nonlinear pulse generation analysed in detail in chapter 7. The regenerative memory operates under the physical principle of excitable regeneration [26], in which a generated pulse in the RTD-PD-LD recirculates indefinitely in an optical fibre delayed-feedback loop enabling robust regenerative signal buffering with the potential to implement memory and logic operations.

This regenerative system follows the strong interest that has been devoted during the past two decades on the effects of communication time delays in biological systems and how they can influence the synchronisation dynamics between distant coupled neurons. For instance, it was shown that dynamical systems mimicking coupled neurons exhibit stable periodic pulsating regimes, instead of a stable steady-state regime. Following the seminal work of Ikeda [18], the concept of pattern memorisation in time delayed bistable systems was addressed in optoelectronic systems.

Departing from these initial works, our work involves an excitable element, the RTD-PD-LD resonator, as the nonlinear node (see chapter 7 for more details about RTD-based excitable system). Because of this crucial difference, our approach benefits from the robustness and self-healing properties that are typical of neural signals that do not exist in bistable systems. Specifically, the excitable response of the nonlinear node guarantees a strong and well-defined all-or-nothing pulse response which is almost identical in shape and duration to any supra-threshold incoming signal, enabling the reshaping and healing functionality of incoming signals. This can be employed in the various key communication buffer functions, which include writing, storage, reshaping/healing and XOR operation.

Figure 8.11 presents a typical example of experimental time traces showing the writing and storage of the regenerative memory using an electrical injection of a complex pattern of 8-bits (11011101). The binary-coded data streams were generated using an arbitrary function generator. For purposes of demonstration and experimental convenience, the data signals were injected electrically, although the signals could be also injected optically, taking advantage of the optical input port of the RTD-PD. By operating the RTD-OEO in an excitable regime (close to the NDC region), we can store and regenerate optical bits of information in the fibre, the empty region signalling the ‘0’ bits and the excitable optoelectronic pulses the ‘1.’ When these bits are reinjected into the RTD-PD, they trigger the generation of a new excitable cycle. This regenerative mechanism occurs after each round trip in the fibre. This is extremely robust because even if the bit sequence is strongly deteriorated, the all-or-none nonlinear response of the excitable RTD-PD enables the signal regeneration. In the experiments, we used a low-loss optical fibre loop with a 46 μs cavity round-trip time, τ. The time τ was chosen to ensure that the memory buffer is much larger than the typical excitable lethargic time, T_l, of the RTD-PD-LD excitable system (in this case, $T_l \sim 500$ ns). The regeneration lasted more than 10^4 round-trips, and was only limited by the acquisition time of the oscilloscope

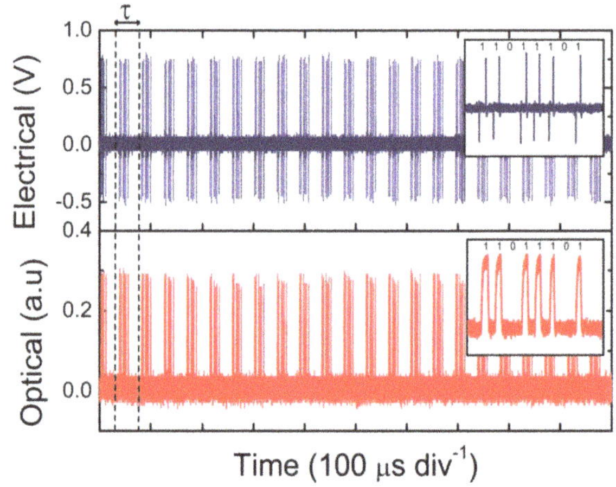

Figure 8.11. Experimental recorded time traces of the regenerative memory showing storage of an 8-bit binary-coded data stream (11011101) in a 46 μs cavity round-trip time. Shown inset is the time trace of a single round trip. Reproduced from [22], CC BY 4.0.

employed. Real time measurements suggest that storage of a data stream can be achieved for several minutes without using any temperature or vibration controllers for both optical fibre and optoelectronic circuits.

It is noteworthy that the writing process of the regenerative memory can be scalable to multi-gigahertz operation, and is only limited by the laser diode frequency response. The information can be written with a single addressing pulse, either using electrical or optical incoming data, and can be stored in standard low-loss single mode silica optical fibres, which is very beneficial for practical applications. In summary, exploiting such a peculiar neuron-inspired excitable electro-optical response using OEO systems is a conceptual breakthrough that has been almost unexplored in the field of optical data communications and could open up new applications, such as clock and timing, optical memories and data buffers.

8.5 Conclusions

In this chapter we have discussed an optoelectronic self-synchronised oscillator based on the integration of an RTD-PD for O/E conversion, a laser diode for E/O conversion and an optical fibre delay line as an optical storage element, described as the RTD-OEO. With this configuration, stable free-running self-locked oscillations with a phase-noise reduction of more than 50 dB at 10 kHz frequency offset from the carrier frequency were achieved without requiring RF or optical amplification. When a dual-loop topology was introduced, the side modes associated with the delay line could be strongly suppressed. The level of the side modes was suppressed by 40 dB, resulting in overall side mode suppression ratio of −60 dBc. The RTD-OEO enables the generation of high-purity microwave signals at multi-gigahertz frequencies.

We have also discussed the implementation of a regenerative photonic memory using the RTD-OEO system. The regenerative RTD-OEO memory operates under the physical principle of excitable regeneration in which a generated neuro-like pulse in the RTD-OEO recirculates indefinitely in an optical fibre delayed-feedback loop, enabling robust regenerative signal buffering. Regenerative RTD-OEOs enable the reshaping and healing of incoming electrical or optical signals, which can be used to implement the key buffer functions required in optical communications, such as writing, storage, reshaping/healing and XOR operation.

Finally, in this chapter we have introduced a numerical model using delayed differential equations based on a time-delayed feedback Liénard oscillator driven by white Gaussian noise and coupled with laser rate equations. This model predicted the experimental behaviour of the RTD-OEO as a function of the time delay and the feedback strength. In addition, once extended to include all the noise contributions of a typical OEO, such as flicker noise from environmental changes in the fibre, this model will enable scientists and engineers to accurately design OEOs based on RTD-PD devices and explore a variety of architectures. Furthermore, the complex phenomena expected in the delayed-feedback system can be explored in various applications, which were not covered in this chapter, namely chaos-controlled methods to stabilise a desired dynamical behaviour [27] and noise-induced stochastic oscillatory behaviour, using the time-delayed feedback as a control parameter [28].

References

[1] Johansson L A and Seeds A J 2000 Millimeter-wave modulated optical signal generation with high spectral purity and wide-locking bandwidth using a fiber-integrated optical injection phase-lock loop *IEEE Photonics Technol. Lett.* **12** 690–2

[2] Mork J, Tromborg B and Mark J 1992 Chaos in semiconductor-lasers with optical feedback —theory and experiment *IEEE J. Quantum Electron.* **28** 93–108

[3] Dahmani B, Hollberg L and Drullinger R 1987 Frequency stabilization od semiconductor-lasers by resonant optical feedback *Opt. Lett.* **12** 876–8

[4] Tang S and Liu J M 2001 Chaotic pulsing and quasi-periodic route to chaos in a semiconductor laser with delayed opto-electronic feedback *IEEE J. Quantum Electron.* **37** 329–36

[5] Yao X S and Maleki L 1996 Optoelectronic oscillator for photonic systems *IEEE J. Quantum Electron.* **32** 1141–9

[6] Gomes N J *et al* 2009 Radio-over-fiber transport for the support of wireless broadband services invited *J. Opt. Netw.* **8** 156–78

[7] Romeira B M P 2012 Dynamics of resonant tunneling diode optoelectronic oscillators *PhD Thesis* (Department of Physics, Universidade Do Algrave)

[8] Maleki L 2011 SOURCESThe optoelectronic oscillator *Nat. Photon.* **5** 728–30

[9] von Wangenheim L 2011 On the Barkhausen and Nyquist stability criteria *Analog Integr. Circuits Signal Process.* **66** 139–41

[10] Lee K H, Kim J Y, Choi W Y, Kamitsuna H, Ida M and Kurishima K 2008 Low-cost optoelectronic self-injection-locked oscillators *IEEE Photonics Technol. Lett.* **20** 1151–3

[11] Shumakher E *et al* 2008 An InP HBT-based oscillator monolithically integrated with a photodiode *J. Lightwave Technol.* **26** 2679–83

[12] Seeds A J 2002 Microwave photonics *IEEE Trans. Microwave Theory Tech.* **50** 877–87

[13] Yao J P 2009 Microwave photonics *J. Lightwave Technol.* **27** 314–35

[14] Fortier T M *et al* 2011 Generation of ultrastable microwaves via optical frequency division *Nat. Photon.* **5** 425–9
[15] Liang W *et al* 2015 Ultralow noise miniature external cavity semiconductor laser *Nat. Commun.* **6** 7871
[16] Yao X S and Maleki L 2000 Multiloop optoelectronic oscillator *IEEE J. Quantum Electron.* **36** 79–84
[17] Erneux T 2009 *Applied Delayed Differential Equations* (Berlin: Springer Science + Business Media)
[18] Ikeda K, Kondo K and Akimoto O 1982 Successive higher-harmonic bifuractions in systems with delayed feedback *Phys. Rev. Lett.* **49** 1467–70
[19] Murphy T E *et al* 2010 Complex dynamics and synchronization of delayed-feedback nonlinear oscillators *Philos. Trans. Royal Soc. A: Math. Phys. Eng. Sci.* **368** 343–66
[20] Illing L, Gauthier D J and Roy R 2007 Controlling optical chaos, spatio-temporal dynamics, and patterns *Advances in Atomic Molecular and Optical Physics* (Advances in Atomic Molecular and Optical Physics vol 54) ed P R Berman, C C Lin and E Arimondo (San Diego, CA: Academic) pp 615–97
[21] Kouomou Y C, Colet P, Larger L and Gastaud N 2005 Chaotic breathers in delayed electro-optical systems *Phys. Rev. Lett.* **95** 203903
[22] Romeira B, Javaloyes J, Figueiredo J M L, Ironside C N, Cantu H I and Kelly A E 2013 Delayed feedback dynamics of Lienard-type resonant tunneling-photo-detector optoelectronic oscillators *IEEE J. Quantum Electron.* **49** 31–42
[23] Figueiredo J M L, Romeira B, Slight T J, Wang L, Wasige E and Ironside C N 2008 Self-oscillation and period adding from resonant tunnelling diode-laser diode circuit *Electron. Lett.* **44** 876–U207
[24] Slight T J, Romeira B, Wang L Q, Figueiredo J M L, Wasige E and Ironside C N 2008 A Lienard oscillator resonant tunnelling diode-laser diode hybrid integrated circuit: model and experiment *IEEE J. Quantum Electron.* **44** 1158–63
[25] Callan K E, Illing L, Gao Z, Gauthier D J and Scholl E 2010 Broadband chaos generated by an optoelectronic oscillator *Phys. Rev. Lett.* **104** 113901
[26] Romeira B, Avo R, Figueiredo J M L, Barland S and Javaloyes J 2016 Regenerative memory in time-delayed neuromorphic photonic resonators *Sci. Rep.* **6** 19510
[27] Blakely J N, Illing L and Gauthier D J 2004 Controlling fast chaos in delay dynamical systems *Phys. Rev. Lett.* **92** 193901
[28] Balanov A G, Janson N B and Scholl E 2004 Control of noise-induced oscillations by delayed feedback *Phys. D: Nonlinear Phenom.* **199** 1–12

Chapter 9

RTD photonics for neuromorphic computing

Chapter 9 deals with artificial neuron RTD-based photonic chips and their relevance in emulating brain activity and regenerative memory observed in biological neuronal systems in the context of neuromorphic spike-based information processing and computing applications. It covers single and autaptic (self-delayed) RTD-based neurons and networks, including all-or-none spike response, high-speed optically-induced spike generation, mixed-mode oscillations (MMOs), spike-based data encoding, storage and signal regeneration, and the realisation of feedforward spiking neural networks (SNNs).

9.1 Introduction

Future information and communication technologies need power efficient and ultrafast brain-like (neuromorphic) systems due to the exponential expansion of artificial intelligence (AI) applications. Scientists are just beginning to develop neuromorphic chips, the hardware of an artificial brain, while AI systems using conventional central processing units (the computer) and neural network algorithms are evolving rapidly. In this chapter, we present the current state of neuromorphic photonic circuits based on solid-state optoelectronic oscillators made of nanoscale double barrier quantum well (DBQW) resonant tunnelling diodes (RTDs). In chapter 7, we introduced the key nonlinear dynamical properties of RTD photonic devices. This chapter discusses artificial solid-state neuron RTD-based photonic chips with and without delayed perturbations in the context of neuromorphic applications, and their relevance in emulating brain activity and regenerative memory observed in biological neuronal systems. The latest studies on excitable and delay dynamic properties of single and autaptic (delayed) artificial RTD photonic neurons, including all-or-none response, spike-based data encoding, storage, signal regeneration and signal healing, are covered in this chapter. A brief discussion of the realisation of feedforward SNNs using RTD-powered nodes is also

doi:10.1088/978-0-7503-5714-2ch9

covered in this chapter. The results discussed here will enable the development of high-performance optoelectronic devices in high-speed brain-inspired circuitry and optical memories for neuromorphic computing.

9.2 Neural dynamics in artificial RTD neurons

The biological neuron, as shown in figure 9.1(a), consists of a soma region (cell body nucleus), dendrites (thin structures that arise from the cell body) and axon terminals (a long cellular extension that arises from the cell body). The neuron's soma in the large neural networks of mammalian and human brains produces action potentials (electrical spikes) with amplitudes in the μV–mV range and durations of 0.1–1 ms. The spikes travel across the axon to the next neurons through synapses, requiring less than 10 fJ per spike. Self-connections (recurrent links) within the same neuron have also been identified.

The synapses, which are 20–40 nm wide gaps between the axon end and the dendrites, transmit the signal either chemically by releasing neurotransmitters or electrically, depending on the type of the synapse. The typical performance of a biological neuron is described in figure 9.1(a).

The artificial RTD-based neuron consists of a neuromorphic optoelectronic microchip formed by a nanoscale DBQW resonant tunnelling diode, which can be either integrated with a photodetector (RTD-PD) or a laser diode (RTD-LD) source (or both), see figure 9.1(b), providing both electrical and optical action potentials (spikes). The typical active area of the RTD component described in this chapter is larger than 10 μm^2 and the usual performance of an artificial RTD photonic neuron of such dimension is displayed in figure 9.1(b). Although the footprint of devices is much larger than the nanoscale dimensions of the DBQW nanostructure (around 10 nm), the RTD-based neuromorphic integrated system could, in principle, be reduced to dimensions (excluding the electrical contacts) of only a few micrometres (or even submicrometer), taking advantage of recent advances in the nanofabrication of nano-photodiodes and nanoscale light sources (nanoLEDs and nanolasers) with wavelength and sub-wavelength scale dimensions [1].

Furthermore, several groups have already reported on high-performance RTD devices with active areas well below 10 μm^2. This would bring several advantages because the energy per spike of the artificial neuron could be substantially reduced ($\ll$1 pJ). In the neuromorphic integrated system, the RTD component provides a non-monotonic current–voltage (I–V) characteristic with a nonlinear region of negative differential conductance (NDC) (figure 9.1(c)). Depending on the direct current (DC) bias point, i.e., the intersection point between the load line and the nonlinear I–V curve, the device can be operated in various dynamical regimes relevant to the neuromorphic application, such as bistability, monostable and self-sustained oscillations, and excitable dynamics.

9.2.1 Bistability

The load line typically intersects the nonlinear I–V curve in one of its two positive slopes, also called positive differential conductance (PDC) regions (I and III in

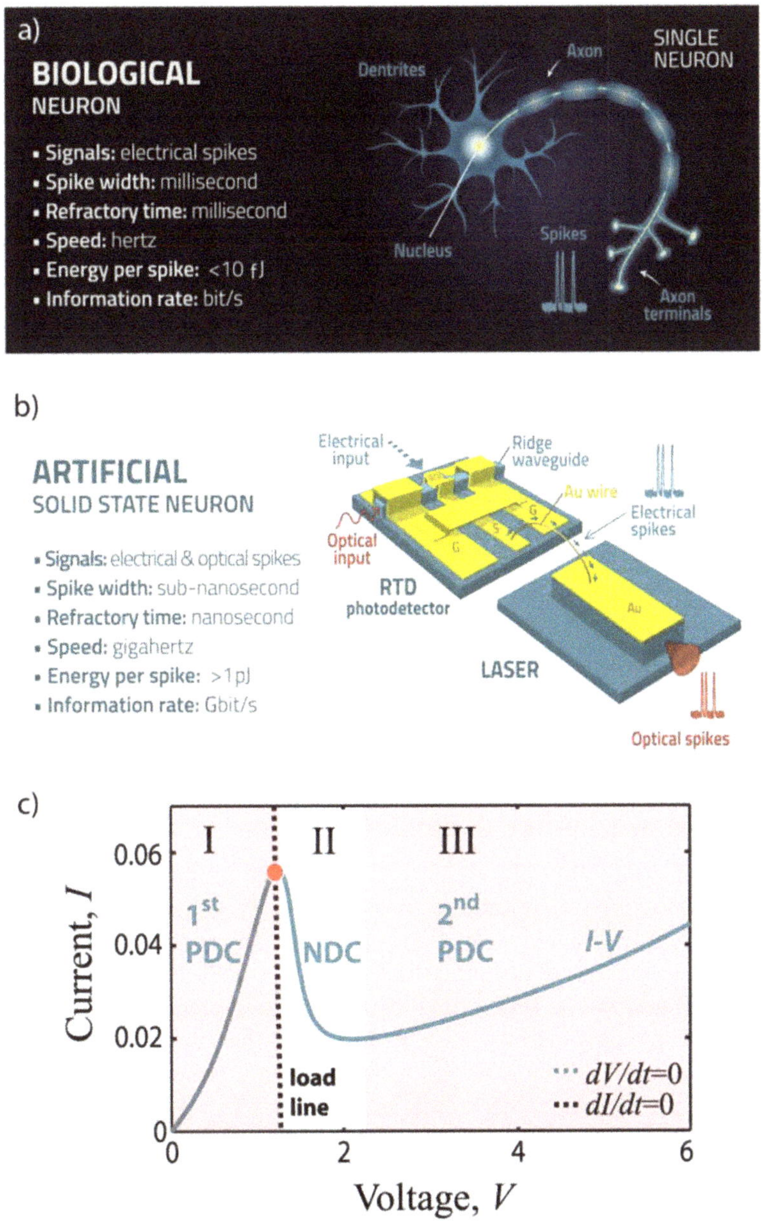

Figure 9.1. Neural dynamics in artificial RTD neurons. (a) Artistic view of a biological neuron. (b) The solid-state neuron microchip consists of an RTD-photodetector connected in series with a semiconductor laser diode using a gold (Au) wire bond—the RTD-LD circuit (chapters 5 and 6). Both electrical and optical high-speed excitable spike signals can be activated, either electrically or optically. (c) N-shaped nonlinear current–voltage characteristic (solid curve) of a resonant tunnelling diode showing the regions of negative (II) and positive (I and III) differential conductance. The load line is also shown (dashed curve), which defines the operating point of the RTD-based neuromorphic optoelectronic circuit. Reproduced from [2]. CC BY 4.0.

figure 9.1(c)). Within either of the two PDCs, the device remains at a steady state but bistability is achieved if the load line intersects both PDCs. In this bistable regime, the noise stemming from, for example, thermal fluctuations in the circuit can induce transitions between the two stable states and a phenomenon of stochastic resonance is observed (as discussed in chapter 7, section 7.3).

9.2.2 Monostability and self-sustained oscillations

In the NDC, region II in figure 9.1(c), the system operates as a high-frequency nonlinear self-sustained oscillator. In this situation, the quality of the high-speed radio-frequency signals can benefit from the high-quality factor of the delayed optoelectronic feedback, enabling ultra-low phase noise self-sustained oscillations (chapters 7 and 8).

9.2.3 Excitable dynamics

When the DC bias point is set at the border between the first PDC and the NDC, region I in figure 9.1(c), neural, type II, excitable dynamics are achieved (sections 9.3.1 and 9.3.2). Multi-pulse excitable bursting is also obtained, such as when biased at the border between the NDC and the second PDC, region III in figure 9.1(c), as a result of a bistability condition that can also occur in this region. Depending on the perturbation and load line, periodic and aperiodic MMOs can be obtained under a proper external modulation (see section 9.3.3). In the excitable dynamic regime, the delayed optoelectronic feedback leads to self-regeneration of the fired excitable pulses enabling regenerative memory operation, as discussed in detail in section 9.3.4.

9.3 Excitable neuron-like dynamics

Neurons exhibit excitability [2], which is the dynamic property that is essential for biologically inspired AI. Excitability is a concept that was originally coined to describe the ability of living organisms, such as nerves or neurons, to respond strongly to a weak external stimulus that exceeds a well-defined threshold. If the system is perturbed from its rest state, it can return to its steady state in two ways. If the perturbation remains below a specific threshold, then the relaxation is exponential. Above this threshold, the system has to perform a large orbit that involves the whole phase-space topology before relaxing again towards the unique fixed point (the rest state). Such a relaxation consisting of two widely different transient regimes towards a unique attractor defines the so-called excitability phenomenon (typically with a pulsed shape). Notably, this unique pulse (or spike) response depends only on the characteristics of the excitable system at hand and not on the details of the stimulus. During its large excursion in the phase space, the system cannot respond to another perturbation, which defines the so-called lethargic time (or refractory time), as the temporal extent of the orbit. This refractory period, which is well recognised in physiology, correlates physically to the high quantity of energy released during the excitable response and can be regarded as the time that the system needs to recover before it can produce another response. This period of time occurs in neurons during the repolarisation and hyperpolarisation of the membrane potential.

Due to their intrinsic capability of signal reshaping, the distinctive characteristics of the excitable response confer a high degree of robustness to data transmission and processing systems based on this mechanism. Early works proposed RTD-based devices as a nanoelectronic candidate for cellular neural networks as a form of threshold logical gates for neuron-like processing [3]. Seeking an alternative architecture that supports spikes as information carriers, here we look at DBQW-RTD circuits as excitable neuromorphic spike generators. In the context of photonics, reports on optoelectronic RTDs integrated with photosensitive absorption layers and lasers have shown potential for bio-neural functionalities. Figure 9.2 summarises the key experimental works published in the last decade that demonstrate significant nonlinear neural dynamic functions of optoelectronic microRTDs. The most notable examples include using bistable RTD photodetectors for light-induced stochastic resonance phenomena (figure 9.2(a)) [4], excitability in photosensitive RTD-laser hybrid integrated circuit devices for spiking and bursting (figure 9.2(b)) [5], autaptic (self-feedback) neuron-like signalling (figure 9.2(c)) [6] and mixed-mode neuron-like oscillations (figure 9.2(d)) [2]. Recently, light-induced stochastic spiking was demonstrated in RTD photodiodes (figure 9.2(e)) [7] and in-device coincidence detection (logical AND) and exclusive logical OR (XOR) tasks were demonstrated using an opto-electro-optical (O/E/O) artificial neuron built with an RTD coupled to a photodetector as a receiver and a vertical cavity surface emitting laser as a transmitter (figure 9.2(f)) [8].

In what follows, we cover the recent results of the excitable and delay dynamic properties of single and autaptic (self-delayed) RTD-based solid-state neurons, including all-or-none spike response (section 9.3.1), high-speed optically-induced spike generation (section 9.3.2), MMOs (section 9.3.3) and spike-based data encoding, storage, and signal regeneration using an autaptic artificial neuron configuration (section 9.3.4).

9.3.1 Spiking and bursting

To operate the neuromorphic RTD-LD microchip of figure 9.1(b) (see also chapters 6 and 7) as an excitable system, the optoelectronic circuit is externally activated using either light or electrical signals and DC biased in a monostable condition (below or above the NDC region), i.e., in a unique fixed point at one of the PDC regions. Under these conditions, the circuit is highly sensitive to the external input (which can be pulsed, square or noise), emitting fixed-shape spikes (or bursts of spikes), depending on the DC bias point. Coherence resonance (CR) phenomenon, i.e., a maximum of temporal regularity in the output for a limited noise level, can be activated by using a noise perturbation as the input signal.

Figure 9.3(a) presents experimental results showing the occurrence of spiking and CR behaviour in a neuromorphic RTD-LD circuit activated by electrical noise. At low noise levels, the output is steady. However, for a given noise threshold perturbation, the circuit is perturbed from its rest state electrical and laser spiking outputs (V and S, respectively) are obtained, panels (i) and (ii). The spike intervals appear to be random and follow the stochastic response of noise. Some examples of

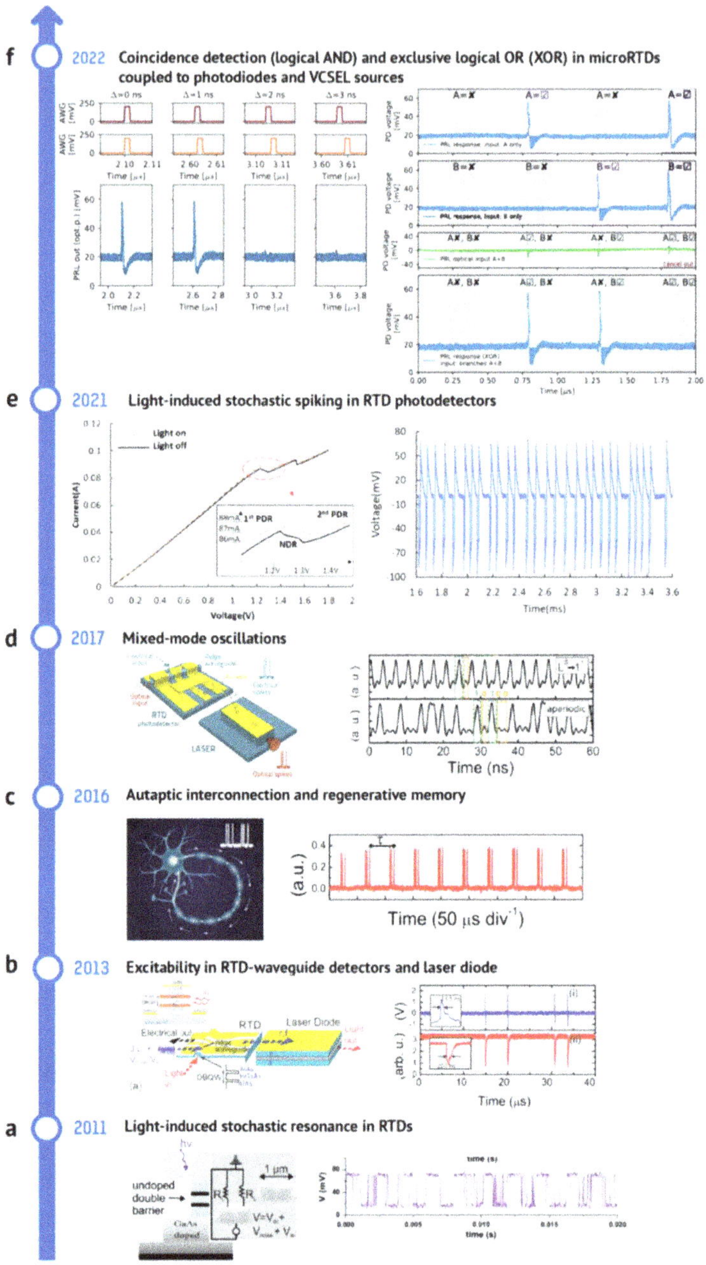

Figure 9.2. Experimental neuron-like dynamical functions and tasks demonstrated in optoelectronic micro-scale RTD-based chips over the last decade. (a) Light-induced stochastic resonances, (b) electrical and optical neuron-like spiking and bursting generation, (c) autaptic (self-feedback) photonic neuron regenerative memory, (d) optical MMOs, (e) light-induced stochastic spiking demonstrated in RTD photodiodes, and (f) demonstration of in-device coincidence detection (logical AND), (right), and exclusive logical OR (XOR) tasks (left). Reproduced from [1]. CC BY 4.0.

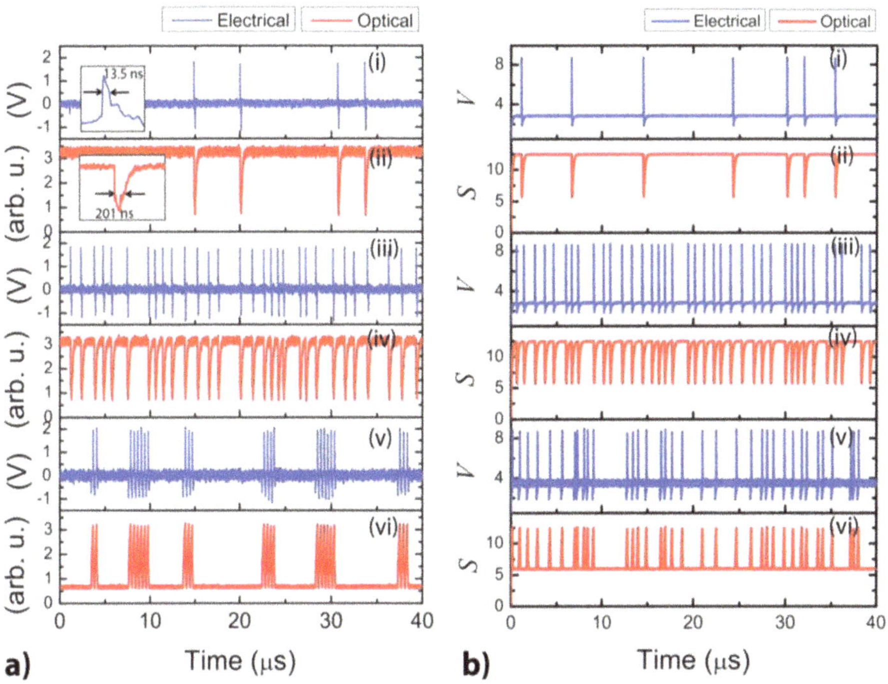

Figure 9.3. Spiking and bursting in RTD-based photonic artificial neuron. (a) Experimental time traces of electrically noise-induced neuron-like pulsing behaviour in an RTD-LD excitable optoelectronic system in both the electrical and the optical domains. The RTD-LD is biased in the first PDC region and is modulated with a noise strength of (i)–(ii) 100 mV; (iii)–(iv) 175 mV. Multipulsing bursts when the RTD-LD is biased in the second PDC region and is modulated with a noise strength of 150 mV (v)–(vi). (b) Numerical simulation of voltage and photon density (V, S) showing noise-induced spike dynamic regimes (i)–(iv) in the first PDC, and (v)–(vi) in the second PDC. The dimensionless noise strength, χ, employed in the simulations: (i)–(ii) $\chi = 0.128$; (iii)–(iv) $\chi = 0.158$; and (v)–(vi) $\chi = 0.310$. Reproduced from [5]. CC BY 4.0.

electrical and optical pulse events are shown in the inset. The voltage pulse consists of an upward and downward pulse corresponding to a full excursion in the phase space. The upward voltage pulse event shows a full width half maximum (FWHM) of ~13 ns, inset of panel (i). The LD intensity output follows the electrical current switching induced by the RTD with a sequence of downward pulses of identical shape and typical FWHM of ~200 ns, inset of panel (ii). At moderate noise levels, panels (iii)–(iv), spike firing is more regular and the interspike intervals (ISIs) are similar, demonstrating signatures of CR behaviour. The ISI statistics of fired spikes allow us to estimate the refractory time at 500 ns. Interestingly, in the first PDC, the response consists of a single isolated peak. But in the second PDC, the temporal response is composed by a succession (or burst) of peaks well separated by the lethargic time. Figure 9.3(a), panels (v) and (vi), shows multipulsing 'bursting' behaviour when the circuit is DC biased in the second PDC region. Since the laser follows RTD-LD-induced electrical current switching, the optical pulses are fired upward.

These observations can be fully explained with the Liénard system model of equations (7.8)–(7.11) (chapter 7, section 7.3) accounting for the current and voltage across the RTD-LD circuit. Figure 9.3(b) illustrates that numerical simulations agree with the experimental electrical and optical outputs. The multiple bursting in the second PDC is also predicted by the model.

9.3.2 High-speed optically-induced spike generation

Excitability and spiking can be achieved at much faster speeds as a result of modifications of the neuromorphic optoelectronic RTD-LD circuit. Using a short wire bond line (transmission line, $L \sim 8$ nH) to connect the RTD and the LD die chips, the lethargic time can be substantially reduced. Figures 9.4(a) and (b) show noise-induced downward and upward optical pulses, respectively, measured in the LD output. The excitation events were also triggered optically, panel (b), by coupling a light signal at 1550 nm to the RTD-PD ridge waveguide using a lensed optical fibre (see chapter 7). The optical signal was amplitude modulated (AM) with electrical white noise. The RTD-PD responsivity was ~0.2 A W^{-1} when DC biased

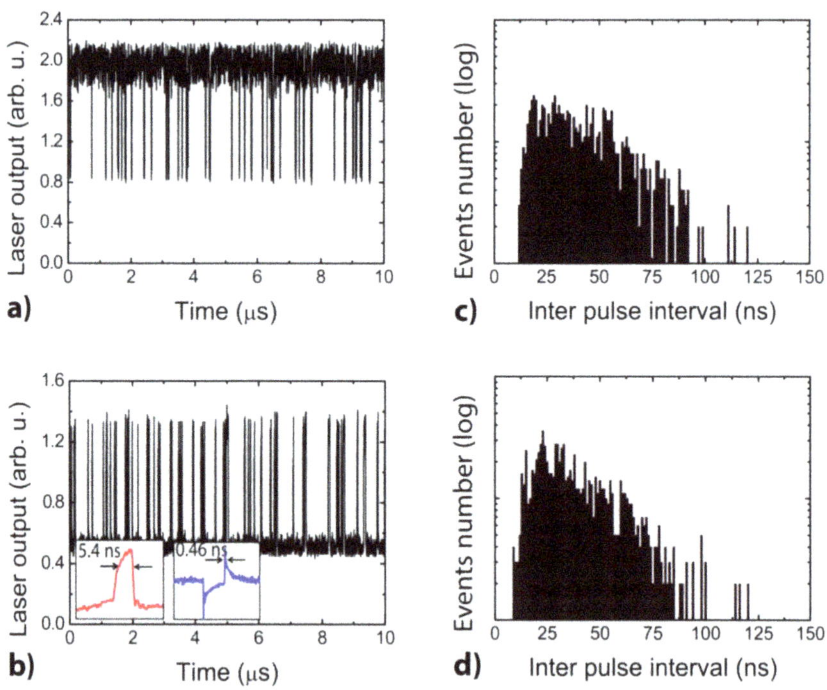

Figure 9.4. High-speed optically-induced spike generation. (a) Electrically noise-induced pulsing in the first PDC using a noise amplitude level of 600 mV; (b) optically-induced pulsing in the second PDC using an optical power signal of 5.5 mW at ~1550 nm and AM modulated with an electrical noise signal with 1.5 V amplitude. Inset: optical and electrical single pulses. Statistic of the times between minima/maxima in the laser output (histogram bin size of 500 ps) when the RTD-LD is biased: (c) in the first PDC region, and (d) in the second PDC region. Reproduced from [5], CC BY 4.0.

in the second PDC. The inset of figure 9.4(b) depicts a single fired pulse event in both electrical and optical outputs. The optical pulse shows a FWHM of 5.4 ns, and the upward voltage pulse event shows a FWHM of around 0.46 ns. Figures 9.4(c) and (d) present the corresponding ISI distribution. This analysis allows us to estimate a refractory time of 12 ns in the first PDC region (figure 9.4(c)) and 9 ns in the second PDC region (figure 9.4(d)). The lethargic time is 55-fold shorter than the lethargic time measured in the previous neuromorphic circuit (~500 ns). A further speed increase is expected with improvements of the microchips, which include reducing the RTD-PD active area (thus reducing the intrinsic capacitance <1 pF) and reducing the series equivalent inductance, mainly determined by the length of the gold bond wires used in the electrical connections.

9.3.3 Mixed-mode oscillations

MMOs describe trajectories that combine small-amplitude oscillations and large amplitude oscillations of relaxation type, both recurring in an alternating manner, as compared with the self-sustained relaxation oscillations. MMOs arise in the context of slow-fast systems, similar to our RTD-based neuromorphic system. The small oscillations arise during the passage of the trajectories near a fold, due to the presence of a so-called folded singularity. The dynamics near the folded singularity are transient, yet recurrent, i.e., the trajectories return to the neighbourhood of the folded singularity through a global return mechanism.

To understand the dynamics of the MMO pattern generation in the case of our neuromorphic optoelectronic circuit, we discuss the current–voltage phase-space diagram displayed in figure 9.5 (left-hand panel). We assume the load line intersects the *I–V* in the first PDC region (rest state), sufficiently close to the NDC region where self-sustained relaxation oscillations can occur. By choosing a periodic sinusoidal modulation, when the driving amplitude is below a given level of amplitude, V_{ac}, the system is only perturbed around the fixed point, as shown in the upper left-hand corner of the phase-space diagram. This corresponds to the small-amplitude oscillations (binary '0'). If we increase the driving amplitude above a threshold value, then oscillations of large amplitude (binary '1') can occur, corresponding to a large excursion in the limit cycle. Using this mechanism, MMO patterns of large and small (LS) amplitude oscillations in the electrical domain are achieved. The two-state level operation can be associated with a binary encoding (0, 1). The LS electrical current patterns directly modulate the laser diode, enabling identical MMO dynamics to occur in the LD intensity optical output.

Figure 9.5 (right-hand panel) shows experimental time traces of several MMO patterns using a neuromorphic RTD-LD circuit. To operate the circuit as a two-state level MMO pattern generator, the bias point was selected in the second PDC region. Depending on the driving frequency and amplitude of the external control AC signal (assuming a fixed DC bias point and circuit parameters), either periodic or aperiodic intermittent MMO patterns can be triggered. Figure 9.5 (right-hand panel) displays the laser output response of the neuromorphic circuit to sinusoidal

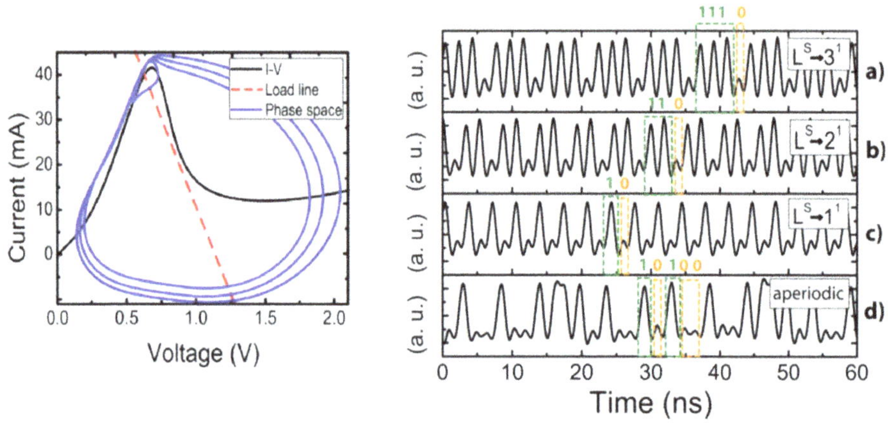

Figure 9.5. MMOs. (Left-hand panel) Nonlinear *I–V* curve, F(V), load line, and numerically simulated phase-space limit cycle of a 3^1 MMO pattern. (Right-hand panel) Experimental time traces of periodic and aperiodic MMOs triggered by an external electrical periodic signal with an amplitude of 281 mV. The different MMO patterns are activated as a function of the injected frequency: (a) 0.544 GHz, (b) 0.564 GHz, (c) 0.586 GHz, and (d) 0.581 GHz. Reproduced from [2], CC BY 4.0.

electrical signals in the range of 0.53 GHz to 0.63 GHz, i.e., close to the natural oscillation frequency of the circuit. The results reveal that increasing the periodic external frequency signal triggers the following periodic MMO sequence: $LS \rightarrow 3^1 \rightarrow 2^1 \rightarrow 1^1$. Between periodic MMOs, complex patterns are observed, including signatures of quasi-periodic and chaotic MMOs. In figure 9.5 (right-hand panel and lower trace), we show an example of an aperiodic sequence displaying random MMOs.

The MMO pattern generation can be used as an efficient switching method to modulate the laser intensity output for applications in pattern and random bit generation. Considering that a small change of intensity or frequency of the external perturbation results in substantial variation of the dynamical characteristics of the MMO patterns, this can have novel applications in data optical encoding.

9.3.4 Regenerative photonic neuron-like memories

This section covers the work on regenerative photonic memories using the autaptic (delayed feedback) configuration introduced in chapter 8. In this setup, an optical delay line connecting both laser and RTD-OD devices (here a low-loss single-mode optical fibre) is placed off-chip, giving a mechanism for the reinjection of the fired excitable optical pulses, emulating an autaptic neuron in an optoelectronic configuration. The regenerative memory operates on the physical principle of excitable regeneration, in which a pulse recirculates indefinitely in an optical fibre delayed feedback loop, allowing for robust regenerative signal buffering and logical processes. The solid-state RTD-based autaptic neuron follows decades of research on how communication time delays in biological systems affect distant coupled neuron synchronisation dynamics. Specifically, the excitable response of the

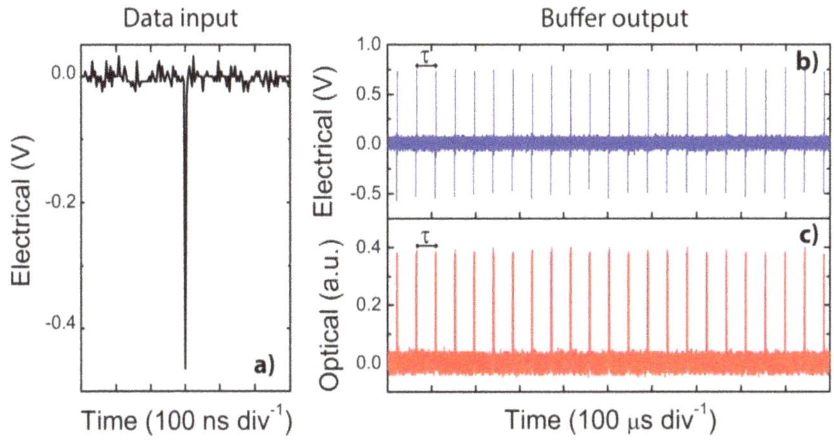

Figure 9.6. Experimental recorded time traces of the regenerative memory output showing writing and storage of a single-bit electrical pulse in a fibre cavity with a round-trip time of τ=46 μs. (Left-hand panel) Single bit data input; (b) electrical and (c) laser photodetected optical outputs. Reproduced from [6], CC BY 4.0.

nonlinear node guarantees a strong and well-defined all-or-nothing pulse response, almost identical in shape and duration to any supra-threshold incoming signal, enabling the reshaping and healing functionality of incoming signals, and playing the role of a memory system.

Figure 9.6 depicts an example of experimental time traces demonstrating the writing and storage process of the RTD-LD regenerative memory utilising a single pulse bit electrical injection, panel (a). We can store and regenerate optical bits of information in the fibre by operating the autaptic neuron in an excitable regime (near the NDC region), with the empty region signalling the '0' bits and the excitable optoelectronic pulses signalling the '1' bits. When these bits are reinjected into the RTD-PD, they cause a new excited cycle to be generated. This regenerative mechanism occurs after each round trip in the fibre and is particularly resilient because, even if the bit sequence is severely deteriorated, the excitable RTD-all-or-none nonlinear response enables signal regeneration.

A wide range of temporal bit patterns were evaluated to determine the robustness of the writing and storage process. Figure 9.7 shows an example of complete regeneration using two-bit (11), panel (a), and four-bit (1101), panel (b), patterns. The bits must be separated by the lethargic time T_l of the excitable system, which eventually defines the maximum bit rate as $1/T_l$. The lethargic time induces a repulsion between the nearest data bits when they get too close, thereby ensuring signal integrity. Moreover, by choosing a bit sequence where the amplitude of the bits is not evenly distributed, panel (c), the system can perform single pass healing by restoring and self-adjusting the received bits to a fixed amplitude. Thus, the unique excitable response of the neuromorphic RTD-LD system enables the implementation of novel forms of regenerative memories that are almost insensitive (in a certain range) to the exact shape or amplitude of the incoming data signals. It is worth

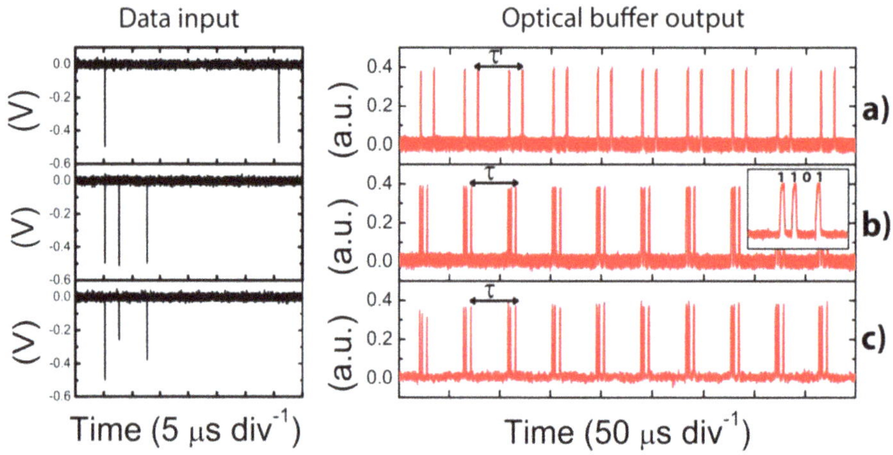

Figure 9.7. Experimental time traces of the regenerative memory output showing writing and storage of binary-coded data streams in a fibre cavity with a round-trip time of $\tau = 46$ μs. (Left-hand panel) Data streams input, and (right-hand panel) laser photodetected optical buffer output. Sequence of (a) 2-bit (11), (b) 4-bit (1101), and (c) and 4-bit (1101) binary-coded data streams. Reproduced from [6], CC BY 4.0.

noting that the regenerative memory writing method is scalable to multi-gigahertz operation, with the only limitation being the laser diode frequency response. The data can be written with a single addressing pulse, using either electrical or optical incoming data, and stored in low-loss single mode silica optical fibres, which is particularly beneficial for practical applications.

9.4 Spiking neural networks

The propagation and processing of information via spikes, inspired by the brain, is one of the main distinguishing features of neuromorphic platforms compared to alternative methods for hardware-based AI acceleration. SNNs [9], as a third generation of artificial neural networks (ANNs), hold significant promise for their further development, particularly due to their prospects for lower energy requirements arising from the intrinsic data representation sparsity in SNNs, as well as their use of the temporal domain for computation and data representation. Because the RTD-based optoelectronic devices investigated in this chapter naturally exhibit excitable spiking responses due to their highly nonlinear characteristics, they are uniquely suitable candidates for light-enabled hardware realisations of spiking neurons and, with additional scaling-up strategies, entire photonic SNNs.

In the previous subsections, we have shown that an RTD photonic neuron can operate as an artificial spiking neuron. However, for more complex data processing, and the realisation of advanced functional tasks, it is essential to go beyond single neuron systems and implement network architectures built from multiple interconnected nodes. In a recent study [10], we have numerically investigated the spike-based information processing capabilities using multiple interconnected (in feedforward architecture) RTD-based optoelectronic spiking nodes where the following

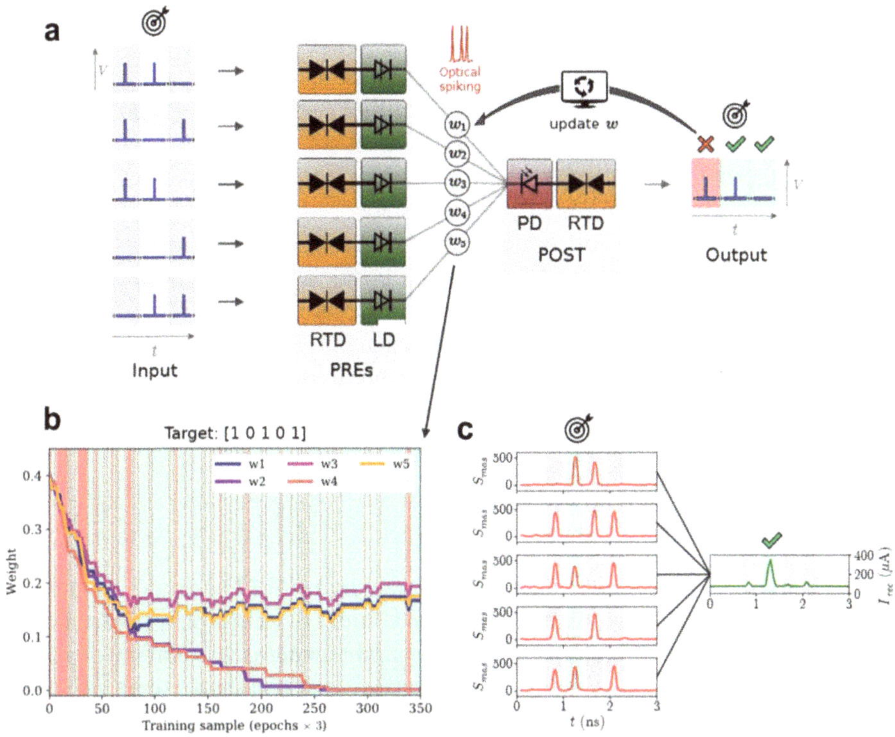

Figure 9.8. Information processing tasks performed with an RTD-powered network. (a) For the classification of spike-based spatiotemporal binary patterns, a 5-to-1 feedforward SNN is used. The system can be trained using an off-chip, supervised, spike-timing-based learning rule (b), and it is capable of conducting spike-based spatiotemporal pattern identification after training (c). Reprinted figure with permission from [10]. Copyright (2022) by the American Physical Society.

interconnected system was considered: (i) master nodes, realising the electrical-to-optical (E/O) link by means of an RTD element coupled to a laser; and (ii) a receiver node, realising the optical-to-electrical (O/E) functionality via an RTD-PD with an incorporated photodetecting input.

To investigate this network, a 5-to-1 feedforward SNN build with nanoRTDs SNN was analysed numerically (figure 9.8(a)). This comprises five input layer master (E/O) nodes and a single output layer receiver (O/E) node. All of the upstream (PRE) nodes are linked to the output (POST) node via unidirectional links, e.g., waveguides, with a weighting function, in this case optical intensity level, taken into account. The PRE nodes encode five binary data streams, resulting in input spatiotemporal spiking patterns that are processed (integrated) in the POST downstream node. The latter only fires a spike when a specific (target) spatiotemporal input pattern is identified, otherwise remaining silent. In this network, the temporal separation between each 5-bit input pattern is set to 420 ps, corresponding to a total network processing capacity of 11.9 Gbps, with prospects for faster processing speeds, via increased parallelisation (higher number of parallel nodes) or faster input sequences.

Similarly to traditional ANNs, the system must initially go through a learning phase (training) before doing inference (recognition) on incoming data. A different solution for network weight modifications is necessary because SNNs are not well suited to traditional error backpropagation training techniques. By introducing a supervised, off-chip, spike-timing dependent network learning rule, an RTD-based spiking network model can be demonstrated (figure 9.8(b)). To train the network, the rule uses labels on input data and temporal separation between pre- and postsynaptic spiking activity of the downstream node. This method was used to train the aforementioned 5-to-1 network, which achieved 94%+ accuracy during the subsequent inference phase on the spatiotemporal patterns (figure 9.8 (c)). The simulation studies show the development of a functional feedforward SNN that computes with spikes and can be potentially realised in hardware with neural nodes based on RTD photonic devices.

9.5 Conclusions

In this chapter, we discussed artificial solid-state neuron RTD-based photonic chips with and without delayed perturbations in the context of neuromorphic applications and their relevance in emulating brain activity and regenerative memory observed in biological neural neuronal systems. We covered recent results on the excitable and delay dynamic properties of single and autaptic (self-delayed) RTD-based solid-state neurons, including all-or-none spike response, high-speed optically-induced spike generation, MMOs, spike-based data encoding, storage and signal regeneration using an autaptic artificial neuron configuration, and the realisation of feedforward SNNs using RTD-powered nodes.

References

[1] Romeira B *et al* 2023 Brain-inspired nanophotonic spike computing: challenges and prospects *Neuromorph. Comput. Eng.* **3** 033001

[2] Romeira B, Figueiredo J M L and Javaloyes J 2017 Delay dynamics of neuromorphic optoelectronic nanoscale resonators: perspectives and applications *Chaos* **27** 114323

[3] Hanggi M and Chua L O 2001 Cellular neural networks based on resonant tunnelling diodes *Int. J. Circuit Theory Appl.* **29** 487–504

[4] Hartmann F, Gammaitoni L, Hofling S, Forchel A and Worschech L 2011 Light-induced stochastic resonance in a nanoscale resonant-tunneling diode *Appl. Phys. Lett.* **98** 242109

[5] Romeira B, Javaloyes J, Ironside C N, Figueiredo J M L, Balle S and Piro O 2013 Excitability and optical pulse generation in semiconductor lasers driven by resonant tunneling diode photo-detectors *Opt. Express* **21** 20931–40

[6] Romeira B, Avo R, Figueiredo J M L, Barland S and Javaloyes J 2016 Regenerative memory in time-delayed neuromorphic photonic resonators *Sci. Rep.* **6** 19510

[7] Zhang W K, Al-Khalidi A, Figueiredo J, Al-Taai Q R A, Wasige E and Hadfield R H 2021 Analysis of excitability in resonant tunneling diode-photodetectors *Nanomaterials* **11** 1590

[8] Hejda M *et al* 2023 Artificial optoelectronic spiking neuron based on a resonant tunnelling diode coupled to a vertical cavity surface emitting laser *Nanophotonics* **12** 857–67

[9] Hendy H and Merkel C 2022 Review of spike-based neuromorphic computing for brain-inspired vision: biology, algorithms, and hardware *J. Electron. Imaging* Review **31** 010901
[10] Hejda M *et al* 2022 Resonant tunneling diode nano-optoelectronic excitable nodes for neuromorphic spike-based information processing *Phys. Rev. Appl.* **17** 024072

IOP Publishing

Charlie Ironside, Bruno Romeira and José Figueiredo

Chapter 10

Conclusion

10.1 Conclusions

In 1928, a very early success of the then-new quantum theory was an explanation of radioactive decay with Gamow's paper 'The quantum theory of nuclear disintegration' [4], which showed how the wave nature of particles explained how energy barriers could be tunnelled through by relatively low energy particle-quantum tunnelling. In contrast, classical physics had no explanation for this behaviour. Since then, quantum tunnelling has been vital to understanding phenomena in chemistry, biology, astronomy and engineering.

This book has presented work on a mesoscopic semiconductor device relying on quantum tunnelling for its operation, the resonant tunnelling diode (RTD), and in particular, has concentrated on how the double barrier RTD can be incorporated into photonics with devices such as the resonant tunnelling diode electro-absorption modulator (RTD-EAM; chapter 4), the resonant tunnelling diode photodetector (RTD-PD; chapter 5), and the resonant tunnelling diode-laser diode (RTD-LD) and resonant tunnelling diode light-emitting diode (RTD-LED; chapter 6). Because the double barrier RTD consists of just a quantum well sandwiched between two thin barriers, it is easily integrated into existing semiconductor optoelectronic components. Furthermore, these RTD enhanced components can be incorporated into systems such as the optoelectronic oscillator (OEO; see chapters 7 and 8) and neuromorphic computing systems (see chapter 9).

Quantum tunnelling is entirely a quantum effect with no classical analogue that can be realised using semiconductor heterostructures and which operates at room temperature. The nonlinear current–voltage characteristics associated with quantum tunnelling can be configured to provide negative differential conductance (NDC), and thereby large electrical gain in high bandwidth (up to ~2 THz) devices. Furthermore, compared to the avalanche photodiodes, this gain is relatively low noise. So far, the bandwidth of the RTD gain has been limited not by the tunnelling

doi:10.1088/978-0-7503-5714-2ch10

process but by the components that make up the circuit that includes the device, such as the electrodes for providing the DC bias.

The RTD-EAM, RTD-PD and RTD-LD all have built-in high bandwidth and low noise gain. This feature that can be utilised to improve existing systems, such as optical communications, and can also be employed in new system concepts, such as the RTD-PD optically controlled 79 GHz RF oscillator for optical to wireless data conversion. In addition, the high bandwidth with gain means that it is straightforward to make these devices oscillate. The oscillation frequencies are determined by the external components and range from a few MHz to up to nearly 2 THz. This unique feature of RTD photonic devices has been exploited in the OEO and for neuromorphic processing and memory.

A brief survey of the history of tunnelling devices reveals that it was realised very early on that, because of the highly nonlinear current–voltage characteristic, they provide an excellent system for the exploration of nonlinear dynamics and the implementation of neuron like behaviour, such as in the FitzHugh–Nagumo model [1] of nerve operation.

The trend in semiconductor electronics is for ever smaller and ever more complicated devices—in 2023, CMOS reached the 7 nm node [2] and 3D chips may be required [3]. At these scales, the basic physics of mesoscopic semiconductor devices – as outlined in chapters 1 and 2—means that it is very likely that an understanding of quantum tunnelling will play an ever more important role as the simple ballistic transport models increasingly become inadequate. The width of the barriers in RTD devices is around 2 nm, so RTDs reached the 2 nm node in the 1990s and a great deal of expertise has been accumulated in how to utilise tunnelling effects to improve and extend photonic devices. Consequently, the future of RTD based devices is well aligned with the dominant trends in semiconductor electronics.

In this book, we have covered RTD photonic devices and their application in a variety of systems. More generally, RTD photonics have been a major influence in the efforts to use quantum tunnelling in semiconductor devices to improve and extend their performance.

References

[1] FitzHugh R 1960 Thresholds and plateaus in the Hodgkin–Huxley nerve equations, *J. Gen. Physiol.* **43** 867–96

[2] WikiChip *7 nm Lithography Process* (https://en.wikichip.org/wiki/14_nm_lithography_process) (accessed July 2023)

[3] Moore S K 2023 Keeping Moore's law going is getting complicated CMOS 2.0 will require exceptional creativity to rewire and 3D-stack the chip, *IEEE Spectrum May* https://spectrum.ieee.org/stco-system-technology-cooptimization

[4] Gamow 1928 The quantum theory of nuclear disintegration *Nature* **122** 805–6

Chapter 11

The Jupyter Python notebooks

11.1 Colab Python notebooks

11.1.1 Introduction

Some of the chapters in this book have associated Python notebooks (with the file extension .pynb). All the notebooks were developed in Google's Colab environment https://colab.research.google.com/notebooks/intro.ipynb.

The optimum way to use these notebooks is to run them on the Google Colab platform—this requires a Google account with the notebooks uploaded to Google Drive and from there on to Colab.

The notebooks are meant to illustrate some of the concepts discussed in the book and can be used as design tools, but they all come with a general note of warning—they only give rough estimates that need to be checked against measurement.

11.1.2 Chapter 2

- **Chpt2 RTD.ipynb:** The Transfer Matrix Method calculates energy dependent transmission probability, T(E), for Al_xGa_{1-x} As RTDs

11.1.3 Chapter 3

- **Chpt 3 I-V fit.ipynb**: RTD Current–voltage and fit to data
- **Chpt 3 Waveguide Modes.ipynb** Refractive index of Al_xGa_{1-x} As and waveguide modes
- **Chpt 3 EAM.ipynb** Electroabsorption and electrorefraction

11.1.4 Web links

The notebooks are available via the following links:

- https://github.com/CodeIron/RTD-Photonics
- https://doi.org/10.5281/zenodo.8026373

www.ingramcontent.com/pod-product-compliance
Ingram Content Group UK Ltd.
Pitfield, Milton Keynes, MK11 3LW, UK
UKHW051852100926
8556IPUK00005B/583